THE CITY IN THE MAKING

THE CITY IN THE MAKING

Marcel Hénaff

Translated by Anne-Marie Feenberg-Dibon

ROWMAN & LITTLEFIELD
INTERNATIONAL
London • New York

Published by Rowman & Littlefield International, Ltd.
Unit A, Whitacre Mews, 26-34 Stannary Street, London SE11 4AB
www.rowmaninternational.com

Rowman & Littlefield International, Ltd. is an affiliate of Rowman & Littlefield
4501 Forbes Boulevard, Suite 200, Lanham, Maryland 20706, USA
With additional offices in Boulder, New York, Toronto (Canada), and London (UK)
www.rowman.com

British Library Cataloguing in Publication Information Available
A catalogue record for this book is available from the British Library

ISBN: HB 978-1-7834-8526-0
ISBN: PB 978-1-7834-8527-7

Library of Congress Cataloging-in-Publication Data

Hénaff, Marcel, author.
[Ville qui vient. English]
The city in the making / Marcel Hénaff ; translated by Anne-Marie Feenberg-Dibon.
pages cm
"Originally published under the title La ville qui vient."
Includes bibliographical references and index.
ISBN 978-1-78348-526-0 (cloth : alk. paper) -- ISBN 978-1-78348-527-7 (pbk. : alk. paper)
-- ISBN 978-1-78348-528-4 (electronic)
1. Sociology, Urban. 2. Cities and towns. I. Title.
HT151.H38413 2016
307.76--dc23
2015025092

∞™ The paper used in this publication meets the minimum requirements of American National Standard for Information Sciences Permanence of Paper for Printed Library Materials, ANSI/NISO Z39.48-1992.

Printed in the United States of America

CONTENTS

OVERTURE

We are told that astronauts traveling on their spaceships in nocturnal zones see the earth's lights as we see the stars: the scintillating archipelagos of Europe or the east and west coasts of North America, the luminous islands of the great metropolises of South America, Africa, or Australia, and the denser ones of Asia, from India to China, from Indonesia to Japan. The inhabited earth looks like an urban galaxy, as if the starry universe surrounding the globe had been projected as a series of cities. It is as if at night humans reflect images of the light they receive back to the cosmos. It is as if, after centuries during which technological civilization fashioned the built environment, we had succeeded in transforming all our artifacts into a second nature, no longer just a series of urban microcosms but a world built on the scale of the planet itself.

However, this is merely an image, no matter how eloquent. Let us come down to earth. As we know, 80 percent of the population in industrialized countries lives in cities today and other countries are following suit. We also know that in half a century there will be practically no truly rural population left. Recent statistics give us the following numbers: the proportion of urban population, *all countries considered*, is 50 percent (that is about 3 out of 6 billion), while in 1800 it was 8 percent; in 1900, 10 percent; and in 1960, 30 percent. There are more city dwellers today than there were humans on the whole planet in 1950, when there were 2.5 billion inhabitants. Today, the urban population increases by about 65 million each year; three-quarters of this increase concerns the megapolises of Asia, Africa, and Latin America, where more than half of the twenty most populated cities in the world are situated. It is tempting to say that the urban form has triumphed everywhere. However, what does this really mean?

The extension of the built environment conforms to the idea of the city only if the word applies to any large collection of living spaces associated with commercial and administrative buildings or leisure. Can we say this without misusing language, or admitting our ignorance of the Millenary evolution of the city, the logic of its emergence and expansion, and also of its present condition?

The question then is as follows. Either the destiny of every civilization is the city, in which case the earth can be considered as an urban planet; or else the growth of the above-mentioned archipelagos of built-up spaces no longer has anything to do with the idea of the city as an organic whole and our familiar image of the world.

If that were the case we would have to face the paradox: *at the very moment when the world seems to become city, the city ceases to be a world.* In short, either the idea of the city has been extended to the whole world and now shapes its organization and image; or else it has been dissolved in spite—or perhaps because—of the extension of the built environment. In that case, we need to consider two hypotheses. Either the dissolution needs to be understood as the beginning of a chaotic evolution, the total loss of any architectural project and requirement for urban planning, which would mean the acknowledgment of an irreversible failure. Or else we are dealing with the emergence of a new paradigm which must be brought to light. Clearly, the stakes in these questions are high, because the answers call not only for understanding the current evolution but also for reflection on urban policy and decisions about the organization of built space.

To find answers to these complex issues, I would like in this short essay to propose the following approach:

1. First, I will analyze the model of the city as a *world*, that is, the city as it has been conceived of since its various origins in numerous cultures as a *monument*, as a vast built environment, or even more as an architectural totality designed to mirror the cosmos. However, we tend to forget that this implies that the city is more than a monument. It is also a *machine* that produces, manages, organizes, and transforms, and as a network comprising road axes, mechanisms regulating the circulation of materials or energy flows, and a place for the exchange of messages and goods. While the two latter dimensions—machine and network—are today more and more obvious, they have existed since the very beginning.

2. Second, I will consider the results of this enquiry in the light of the concept of *public space*, understood in a very general sense. I mean by that an urban site where some buildings such as princely palaces, administrative buildings, fortresses, temples, are clearly distinct from private dwellings and spaces reserved for individual activities. This concept will be contrasted to another one which is central to the Western tradition, in which the phrase, "public space" first of all means a sphere of open communication and free debate whose institutional expression is an elected assembly that decides in open forums the laws of the City, its judicial forms, and its choice of war and peace. This was substantially the Greek model, or at least its ideal, still claimed by Western democracies whose urban spaces were a testimony to that model and still are.

However, we should ask whether the concept of public space can be applied to other civilizations and associated with the monumental visibility of the ancient kingdoms of the Middle East, India, or China, or the states that succeeded them. This is questionable. In speaking about the city, it will therefore be essential to highlight another level of social experience which is more informal and everyday and therefore probably more universal, that of *the common sphere*, understood as a strictly local order of urban relationships, and especially of neighborhoods. These relationships can be random or organized, marked by a variety of attitudes concerning civility or lifestyles, involving the relationships between different genders, generations, professions, linguistic practices, or religious forms. Or they can be linked to dress or food preferences or other expressions of local particularities. In short, they form a *vernacular order* the practices of which differ between cities and neighborhoods and even more between cultures.

In any case, this concept seems essential to understand the current evolution and to go beyond an exclusive focus on the opposition between public and private. The *common space,* whose emblem is the street, does not simply lie between those two but crisscrosses and underpins them. It is also different from what we call the social realm or civil society. Its very distant relation to political institutions—more so in some traditions than others—gives the impression that this *common space* manifests a rejection of all public space or a prudent indifference to it. But this would mean interpreting it as a lack, which it is not. It is neither an enlarged private sphere nor a diminished public sphere. It is certainly specific to urban space where *human diversity* has greater opportunity than anywhere else to be recognized and valorized. And so, the whole concept of public space needs to be rethought. It is precisely this concept, canonically accepted in the West, that is being challenged and revised by the new planetary mechanisms of information and intervention networks encompassing today's cities. The workings of these mechanisms challenge the organization of the built environment.

Perhaps this double investigation will give us at least a partial answer to the initial question about the permanence or disappearance of the classical urban model, assuming that it is applicable across very different civilizations. This would enable us to think in terms of a new paradigm to decipher current changes and foresee possible evolutions; in short, to envisage *the City in the making*.

First Approach

Monument, Machine, Network

INTRODUCTION

In his epoch-making work, published in 1966, entitled *The Architecture of the City*,[1] Aldo Rossi forcefully argued that the question of the city can only be properly theorized by stressing the central role of its architectural reality. This was in opposition to the dominant view developed by American city planners[2] who tended to reduce the urban phenomenon to an environmental problem. Even so, Rossi does not ignore the question of the *site*, which he prefers to call the *locus*. This notion includes the environment but integrates it with a concept of the *urban landscape*, which recognizes the importance of stylistic singularity. Rossi's emphasis on the built environment and on the characteristic monumentality of cities has been valuable to decipher specific research on the city.

However, no matter how fruitful Rossi's approach, he tended to forget that the city is also always a way to concentrate people, organize them, enable them to produce, exchange, and make a living. In short, the city is also a powerful engine of technological transformation and therefore this monument is also a *machine*. That is not all: the space of architectural visibility is just as much a

1. A. Rossi, *L'Architettura della città* (Ed. Clup, Marilio: Padova, 1966); translated *The Architecture of the City* (Cambridge, MA: MIT Press, 1982).
2. Mostly the School of Chicago, and especially Robert E. Park, Roderick D. McKenzie, and Ernest W. Burgess; a very innovative school in many respects.

space of circulation, connections, exchanges, and information; in other words ever since its first appearance the city has also been a *network* and even a network of networks. This triple dimension justifies calling it a *world*. A world is not just the image of a totality, a space offered to our gaze, but it is also a populated space. It is a population that lives according to certain rules, that works (often amid suffering), changes its environment (which it sometimes destroys), accumulates experiences, produces knowledge, techniques, and works. Finally, it consists of groups and individuals who circulate, communicate, debate, oppose one other, fight—sometimes to the death, and exchange goods and information.

However, it is not enough to refer to these three essential dimensions; we also need to evaluate in what ways 1) they are indispensable to the understanding of the specifically urban phenomenon; 2) they make it possible to trace transformations over time; and finally, 3) they provide conceptual tools for developing a hypothesis for a new paradigm linked to the changes of the contemporary city.

I

MONUMENT: THE CITY AS TOTALITY AND AS IMAGE OF THE WORLD

According to archaeologists, the appearance of cities goes back ten or even more millennia, which is a lot on the scale of our civilization but very little when compared with the emergence of *Homo sapiens*. The urban phenomenon appears independently in different places on the planet and during very different times. We have no evidence that the cities that first arose in the Fertile Crescent were the origin of the urban phenomenon in the valley of the Indus or in China. It is also clear that the formation of cities in Mesoamerican civilizations (Olmec, Toltec, Mayan, and Aztec) owes nothing to any other civilization outside the Americas. The fact that it is a *sui generis* phenomenon clearly indicates that it is linked to specific conditions that enable its emergence. Nevertheless, it is not a universal occurrence: there are still some civilizations today, although few, that have not experienced it. We need to ask the question again: why cities? Why did people who for tens of millennia had learned to live and survive perfectly well together in many different ways—as hunter-gatherer societies and then much later, as farmers or nomadic cattle breeders (or according to mixed forms such as in itinerant horticulture)—why did they, in some places of the planet, start to live together in restricted spaces, to erect monuments, and build imposing or modest dwellings made of clay, wood, bricks, or stones?

THE CASE OF THE FIRST KNOWN CITIES: THE FERTILE CRESCENT, MESOPOTAMIA

As of today, the oldest and best attested cases of the construction of cities are in Mesopotamia and the Fertile Crescent. High-quality documentation in this area is expanding. It will be quite instructive to look again at some of these archaeological discoveries.[1] We have at our disposal good analytical criteria concerning the material conditions of settlement changes, changes in building techniques, the purposes of buildings, and the social differentiation of the built environment. Let us go over these different points.

All scholars agree on one fact: the urban phenomenon appeared in sedentary agricultural societies. This is not a recent discovery; archaeologists have been saying this since the nineteenth century. However, today we have learned that there have been sedentary hunter-gatherer societies, as was the case in Palestine or in the hills of Syria during the Natufian period (12,500–10,000 BC).[2] Apparently, this situation resulted from an abundance of wild cereals and game, which favored the domestication of plants and animals. This shows that we need to be careful about the use of the concept of settlement living, for it existed before agriculture. Similarly, the concept of domestication should be used with

1. S. N. Kramer, *The Sumerians: Their History, Culture, and Character* (Chicago: University of Chicago Press, 1963); J. L. Huot [ed.], *Préhistoire de la Mésopotamie* (Paris: Ed. du CNRS, 1987). A. Parrot, *Sumer: The Dawn of Art* (New York: Golden Press, 1961); R. C. McAdams, *Heartland of Cities* (Chicago: University of Chicago Press, 1981); O. Aurenche, *La Maison orientale: L'architecture du Proche Orient ancien des origines au milieu du quatrième millénaire* (Paris: Ed. Geuthner, 1981). H. J. Nissen and J. Renger, *Mesopotamien und seine Nachbarn* (Berlin: D. Reimer, 1982); J. C. Margueron, *Les Mésopotamiens* (Paris: A. Colin, 1991).

2. J. Cauvin, *Les premiers villages de Syrie-Palestine du IXe au VIIe millénaire avant Jésus-Christ* (Lyon: Maison de l'Orient, 1978); J. Cauvin, *The Birth of the Gods and the Origins of Agriculture* (Cambridge, UK: Cambridge University Press, 2000).

circumspection since it is not just, as Cauvin,[3] insists, a way of meeting the need for meat consumption, but is also part of changing symbolic relations between humans and animals which gave rise to a new desire to dominate the entire biotope.

Second caveat: agriculture appeared first of all in the hills of the Fertile Crescent, that is, in zones abounding with wild animal or vegetal resources such as sheep and goats, grains, pulses, fruit, and other plants. Nevertheless, urban development did not take place in these hills but in fluvial valleys where such alimentary resources were almost nonexistent. This surprising fact must be explained. Before considering the answers of recent archaeology, it will be useful to note the changing forms of habitat.

First of all, it must be noted that even if the city implies a break with the village, the village habitat is nevertheless the form par excellence of the settlement process and the space where the first decisive morphological transformations of the house took place. The village form existed alone for more than two millennia before the city. Therefore we should not overestimate the break, real though it may be as we will see, that took place between these two types of agglomeration, as several historians are tempted to do by giving excessive importance to foundational rituals. The truth is that the millennia when rural habitats and their building techniques were elaborated, were determining for the transition to urban architecture. Indeed, with respect to the dwellings of hunters-gatherers of the preceding ages, one of the major innovations in habitat is the change from the circular to the *rectangular shape*; furthermore, this new type of building ceases to be underground but on the contrary tries to rise above ground.[4] Cauvin gives the following explanation: "The round house from the start has a defined area for living and, from the perspective of adding extensions, presents an impasse. . . . The construction of rectangular buildings supposes that people knew how to make a free-standing rectangular wall-plan stand up, and also how to fit two walls to-

3. J. Cauvin, 1997, p. 130–32.
4. J. C. Margueron, *Les Mésopotamiens*, p. 9–17.

gether at right angles."[5] Archaeology shows that the straight wall only appeared at first to divide round houses, but then evolved as the most efficient technique for managing and expanding living space. However, in some places, the circular form persists in spite of the well-established mastery of the straight wall technique. According to Cauvin, this persistence is explained by cultural reasons and not by technical inadequacy; in other words, symbolic traditional constraints are no less powerful than the functional constraints that lead to innovation.

This change is significant because the rectangular form makes it possible to have both horizontally and vertically extended buildings. They become flexible and extensible structures, renewable *modules* that can respond to demographic growth. The system of terraces and upper rooms not only opens up additional living spaces but also food storage; terraces especially can more easily be protected from rodents or humidity. This formula was invented in villages, and contains the seed of the extensible orthogonal grid. It was maintained and developed in urban architecture because it turned out to be the most efficient way of organizing interior living spaces, intra-urban traffic routes, and exterior walls or monuments. From that point of view, the morphological continuity is quite significant.

Another remarkable characteristic of agricultural dwellings in the Fertile Crescent is the apparent uniform purpose of the houses: all seem to assume the same basic functions of lodging and food storage. Some constructions seem set apart as attics or storage spaces but there is no way of determining whether they are common buildings or annexes belonging to particular families. Their spatial arrangement tends to favor the second hypothesis. In contrast, there is one characteristic that makes it easier to pinpoint the specificity of urban space right from its first appearances at the turn of the IX–VIII millennium BC: the existence of *collective buildings*. There are different types: exterior walls, temples, palaces, or canals. We should remember this very visible feature and

5. J. Cauvin, *The Birth of the Gods*, 128–29: "The Square House."

ask whether it should be considered essential to the city, that is, the existence of a public space, without yet defining the meaning of the term.

One last remark: in the *village organization* every construction seems to be more or less the same size. This has been considered as the probable sign of a great equality in condition, even though there is no evidence for this. Some houses appear more important than others and we do not know whether we are dealing with the dwelling of a chief or simply one belonging to a wealthy individual. On the contrary, in the case of a *city* there is a clear distinction between the dwellings of ordinary people and the homes of elites whether they be political (palaces of kings and nobles), religious (temples), or social (merchant families or enriched artisans). Thus, although *hierarchical differentiation* in itself is not a new phenomenon in agricultural civilizations, the city is the site of its original visibility discoverable in the architecture itself.

We can now answer our earlier question. The agricultural revolution in the Fertile Crescent took place in the hills, but the cities developed first in the valleys, more specifically on riverbanks. While in the hills larger and larger villages were developing, they nevertheless did not show the distinctive signs of the city. This means that the growth of agricultural resources alone cannot account for the urban phenomenon, contrary to what has often been claimed. Something else was needed: first of all innovation in the use of resources, which implies both material and social conditions. The archaeological data show that this new situation was initially linked to the transfer of grain farming to riverbanks where the soil is fertile and where irrigation canals could make a constant water supply available. Here we have an important fact: the link between shortage and surplus. Shortage means the lack of resources such as wood, construction stone, asphalt, and later, minerals. They had to be imported, which becomes possible when exchanged for surplus agricultural products. From the start, this necessary exchange represents not only a decisive economic change but also a considerable technological and political change resulting from transportation requirements that favor or even trigger them. To quote another specialist in this area:

> How could these exchanges take place while the wheel and the donkey only appear, it seems, around 3000 BC? In the absence of any other means of transportation, except for human carrying, only rivers and canals could provide the necessary infrastructure. This situation means that the first cities in the Mesopotamian environment are closely linked to a river or canal and that this association was of course intended to insure a good water supply for daily life, to allow for the irrigation of farm land but also to maintain relations. It was not so much the watering hole that favored the blossoming of the city as was the *river axis*, in other words, *running water*. Furthermore, it was not only the management of large agricultural domains that favored the expansion of cities, as we thought for so long, but also the development of commerce.[6]

The movement of exchange along different networks is an essential starting point of the urban phenomenon. As we will shortly see, Max Weber had a good understanding of this. Nevertheless, we must note a point that touches on the second set of conditions: social factors. The above-mentioned text refers to large domains, which means that someone's surplus is assured by someone else's labor. We are dealing here with another complex history of the domination of some groups over others. While this inequality is indeed linked to the "agricultural revolution," it cannot be reduced to merely one of its consequences. Slavery in particular is better explained by the capture of war prisoners than by the disparity of resources even though the latter perpetuates it. Perhaps it would better to speak of servants or dependants than of slaves. Documents such as the Hammurabi Code prove that Mesopotamian societies were hierarchical very early on, with inequalities corresponding to the farming system: notables are *awilum*; middle classes; *muskenum* and servants: *wardum*. In Mesopotamia it was

6. J. C. Margueron, *Les Mésopotamiens*, t.2, p. 22.

linked to a very ancient land ownership system, with meticulous legal rules. It clearly corresponded to a system of political domination in the distribution or seizure of land, a practice sanctioned by the city where the king and notables lived. However, to see it as "despotic" would be an error; this is rather a society of debts and services, indeed a society that probably invented the first form of wage labor.[7] Whether there was wage labor or slavery, we still need to understand how the city as built space with an original social organization becomes the mechanism par excellence that mobilizes new forces.

THE URBAN SITE AS SYMBOLIC SPACE AND FOUNDATIONAL RITUALS

The emergence of the first known cities—at least those known today—in the Fertile Crescent and especially in Mesopotamia interests us not so much in order to hypothesize a model of processes that take place elsewhere, but because it signifies a relation between the material conditions of the city and the creation of a symbolic space whose lexicon is formed by those conditions. Whatever the variety of causal combinations that led to its emergence, the city appears as a *monument* from the very beginning. More specifically, it presents itself as a monument that sums up the universe, as the work of men that is also the home of the gods, as the heart of a political space which it expresses, as both its image and instrument.

Rather than recounting an uncertain genealogy, we should limit ourselves to two well-documented elements: foundational rites and representational schemas. These will allow us to envisage two major aspects of the establishment of the city as a monument; on the one hand the constitution of an original symbolic space and on the other hand the formation of a community linked to that space.

---

7. Cf. J. J. Glassner, *La Mésopotamie* (Paris: Belles Lettres, 2002), p. 84 sq.; J. J. Glassner, *Mesopotamian Chronicles* (Leiden & Boston: Brill, 2005).

In every case, one thing stands out in the archaeological evidence or texts from multiple ancient civilizations: the city is the privileged space of the relations between humans and gods; the city inscribes on earth the heavenly world and becomes *the mirror of heaven*. Whether it concerns the sacred mound of the Sumerians, the ziggurat of the Mesopotamians, the Greek or Roman temples erected on hills, everywhere archaeology reveals this one abiding feature (as Kramer has emphasized).[8] Not only does the city from the beginning pertain to religious symbolism, including the citadel and the city walls, but the symbolism is specifically linked to heavenly divinities and to heaven as a whole, to the sun and to light. This is opposed to the world below shown as defeated, as the ancient world of earthly divinities, rural activity, and lineage traditions.

In civilizations as varied as China, Greece, Rome, Japan, the Aztecs, India, and the Middle East, it is obvious that to build a city is to build a world. This is why the city as such is a monument. It is a *global creation* and the site of all particular creations. Research on the first known cities (Palestine, Mesopotamia, the Indus valley, China, and, later on, Egypt) teaches us the following: the city wants to be *the image of the world*. The city is built as a replica of the cosmos, the realization on earth of a world that corresponds to that of the gods. It links and unites those worlds. *Were they not linked before?* Yes, they were, but in other forms which the city made obsolete by utterly transforming the relations of humans to the so-called natural world. The gods or the genies of the fields, of water, of the winds, and forests have fled. The city exiles them to the netherworld and invites into its temples new divine figures projected onto empyrean heavens. These figures have specific names and are often ordered in pantheons which written traditions now begin to celebrate, narrate, and individualize. The city presents itself as a world because it is understood as the expression of forces that are remaking the world. The city opposes the *pro-*

8. S. N. Kramer, *History Begins at Sumer* (Garden City, NY: Anchor Books Doubleday, 1959).

duced world to the former world experienced as *given*, which was the place of an egalitarian alliance and where humans and nonhumans circulated on a common level. The city introduces a tiered separation of heavenly gods, humans, and infernal divinities and reconstructs what it undid according to this new schema. The city is a world because it builds the world and because it simply is the new world.[9]

A Foundational Model: Ancient Rome. The Roman ritual is a good example of a foundational rite that expresses the great solemnity of the situation: the establishment of an urban space is the establishment of the world. According to Roman tradition, the foundation of a city had to follow a very precise ritual, which we know from the *Gromatici Veteres*, a compilation by the surveyors (*agrimensores*) of Trajan's era, the first century BC. This ceremonial of Etruscan origin included four moments:

- *Inauguratio* or reading of heaven (*signum ex coelo*) by the *haruspices* to choose the spot where the city would be established. In the case of Romulus, the sign from heaven was the appearance of twelve vultures; the axis of the world and the center of Rome were marked there vertically.
- *orientatio*, or the choice of geographical axes, north-south/east-west.
- the *limitatio*, or the outline of the ditch circumscribing the space of the city.
- *consecratio*, the final ceremony giving a name to the city and placing it under the protection of one or more divinities.

There is one remarkable point in this ceremony: it is in *heaven* that the location of the city is set. This is the first articulation between the city and the world above. The second is defined by

9. It would be the task of historians of religion to find in the first written evidence everything concerning the increasing importance of this construction model and the urban site as image of the world, and at the same time evaluate the ambiguous relation developing between the figures and traditions of the earlier world, both denigrated as obsolete and venerated as the origin in the figure of the *Golden Age*.

the *orientatio* which determines first of all the east-west axis, called *decumanus*, according to sunrise and sunset and then outlines perpendicularly the north-south axis or *cardo*, that is, the pivot or hinge. This axis supposedly corresponds to an ideal line around which the celestial vault pivots. The space of the city is also linked to the divine world because the *limitatio* or perimeter traced by the plow, the first furrow—*sulcus primogenius*—circumscribes a purified space acceptable to the gods.[10] Within these outlines, the limits of the different parts of the city will be inscribed geometrically on the ground: surrounding walls, streets, and squares. It starts with the outline of orthogonal roads, on both sides of the *cardo* and the *decumanus*. Then comes the choice of the site of the temples, whose first element is the *tumulus* where the Capitoline gods will reign (Jupiter, Juno, and Minerva), in other words the gods of the *caput*, the head of the city from where its entire space can be surveyed in one glance. Below, corresponding exactly to the city center—*umbilicus urbi*—the *mundus* will be dug. This ditch placed in the heart of the city puts the city in contact with the whole world. Earth brought from other regions is buried there and fruits are left as offerings to the infernal gods. Here is the lower world where malevolent forces are confined, thrown into the *mundus*. The double *caput* and *mundus* pole clearly symbolizes the vertical axis of urban space. It looks upward but must also take into account and even integrate its contrary, the ancient world of earth and its obscure forces that are supposed to be dangerous but cannot be ignored.

While the foundational operation is indeed religious, it is also very much legal and geometric. The outlines are made (with the help of a *groma* or measuring stick) according to precise calculations of the lines and angles of an orthogonal system that respects units of magnitude and ensures proper proportions. As stated by Nicomachus of Gerasa, a writer of the first century BC: "In har-

10. A. Charre, *Art et urbanisme* (Paris: PUF, 1983), chapter 1, "Les infrastructures romaines"; P. Grimal, *Roman Cities* (Madison: University of Wisconsin Press, 1983); A. Grandazzi, *The Foundation of Rome: Myth and History* (Ithaca, NY: Cornell University Press, 1997).

mony with the Number, all was created as according to an architectural plan, and also time, movement, the heavens, stars and cycles of all things."[11] The formation of the world was an architectural creation, while vice versa the architectural foundation of the city was the recreation of the world. When cities make their appearance, ancient cosmogonies are rewritten as narratives of divine construction. As such the world itself becomes the first artifact, a product of the divine artisan, like the Greek *demiourgos*.

Two quotes can explain the Roman idea that the City was constructed according to a model of identification with the world: the first, from the time of Augustus, from Ovid: "The limits of foreign territories are well defined; the City of Rome's limits are those of the world—*Romanae spatium est Urbis et Orbis idem.*"[12] This is echoed a few centuries later (in 416 BC) in an observation by the Gallo-Roman Rutilius Namatianus: "Urbem fecisti quod orbis erat": "You made a city that was a world"; to be understood as "you arranged for the city to become the world." The city includes the world, sums it up, integrates it but it does so successfully by first reproducing itself everywhere in identical ways. The world becomes city. In every new province and every country subjected to its might and authority, the Roman conquest repeated the model of the inscription of the city of Rome on the ground of the new site: its sacred space, its gods, monuments, axial division, grids of streets and squares. As the local actualization of the model of the *urbs,* each city became a book of stones, the grammar of spaces teaching Roman ideas. The language and ways of living, knowledge, and the arts would then arrive and prosper. Archaeological excavations on sites of Roman cities in the Po valley, the Rhone valley, those in North Africa or the Middle East, have shown great consistency in the actualization of the model.

Another Foundational Model: Ancient China. The meticulous use of geomancy and calculation make the foundational rites of

---

11. Nicomachus of Gerasa, *The Manual of Harmonics* (Grand Rapids: Phanes Press, 1994) [cited in D. Payot, *Le Philosophe et l'architecte* (Paris: Ed. Payot, 1983), p. 68].

12. Ovid, *Fasti*, book II, 683–684.

the Chinese city rather similar in terms of form to those of the Roman city. Marcel Granet gives a detailed description in his classic work *Chinese Civilization.*[13] The principal source is the *Che King*, a compilation of poetic and ritualistic texts, which complements the *Chou King*, the history book attributed to Confucius. The different phases of the foundation of a city follow rigorous procedures. First comes the identification and determination of the site. The founder, magnificently attired, adorned with jewels and armed with a ceremonial sword, goes on a search for the most appropriate spot to establish the future city. To this end he observes the movement of the sun, the relation between shadow and light, the orientation according to cardinal axes, the direction of streams and winds, the quality of the soil, and the relationships between *ying* and *yang*. On these climatic and geomorphological conditions depends the local emanation—*hsing shih*—of the energy that sustains the world. On the basis of his findings, the founder continues with calculations which then are submitted to divination in tortoise shells. Once all this information is acquired and the religious worthiness of the site has been established through geomancy, the order to build is given when the constellations are favorable at a time following the end of the harvest. The order of construction is precise: it starts with the most sacred element, the rampart. Then comes the temple of the ancestors and the planting of trees whose fruit will be offerings for the ancestors. Construction is done by peasants subject to compulsory labor, following ritual rhythms.

The city perimeter draws a square reproducing the form of the earth itself, which will also be the form of the temples and the lord's palace. The most important and hence sacred elements of the rampart are the doors, placed at the four cardinal points. They are particularly well fortified and protected by rituals designed to heighten their holiness (the most efficient rite being immuring the

13. M. Granet, *Chinese Civilization* (New York: Meridian Books, 1958); see also Wu Hung, *Monumentality in Early Chinese Art and Architecture* (Stanford: Stanford University Press, 1995).

head of a defeated enemy). Granet comments: "The divinity of the town is lodged in the gates and walls."[14] Then follows the construction of altars and temples. The Altar of the Soil can be crude while the Temple of the Ancestors must be magnificent. Then the Lord's residence can be built. Its shape is the same as the city's with a square enclosure and towers. "Square like the town, and surrounded by walls, it is a town in itself."[15]

In imperial cities the symbolism is even more meaningful: the palace is set in the center of the city through which passes the axis of the center of the world—the *ti chung*. The texts are unfailingly precise about this axis: "The spot where the shadow of the culminating point is at one foot and five tenths, designates the center of the earth. It is the place where heaven and earth unite, where the four seasons come together, where wind and rain assemble and where the two principles of male and female are in harmony."[16]

But let us consider the rest of the city. The neighborhoods are divided according to perpendicular lines on both sides of the cardinal axes leading to the gates. The other residences are also surrounded by walls so that streets are circulation strips between closed spaces: "Like the princely residence, the dwellings of the great families are a town in themselves, a town grouped around the great hall where the chief of the family receives the homage of his relatives, turning towards the south."[17] We are not told about the houses of modest families, just as we are not told of those in Babylon, Athens, or Rome. The reason might be that they seem part of a mass rather than monumental. Repressed, the popular city will reemerge vigorously in the industrial city. Of course, this is an entirely different story.

This elementary reminder of the foundational ritual and the construction model of various buildings allows us to measure the exceptional character of urban space in the Chinese tradition.

---

14. M. Granet, *Chinese Civilization*, p. 239.
15. Ibid., p. 245.
16. M. Granet, *Le Tcheou-Li, ou Rites des Tcheou*, translated by F. E. Biot (Paris: Imprimerie Nationale, 1851), vol. 1, p. 201.
17. M. Granet, *Chinese Civilization,* p. 245.

Granet notes the degree to which the city, as opposed to the countryside, tends to monopolize the holiness of the soil, to claim the presence of the ancestors, and to focus the power of the Lord. Rural space is despised and the inhabitants are considered boors, while the city is revered as noble, the site par excellence of good manners and great virtues. In fact, there was a slow evolution that transferred the holy sites of rural traditions to the cities. "Great festivals which were also fairs were held in the Holy Places: there one communed with one's native soil; there one invited one's ancestors to come and be reincarnated."[18] However, little by little "the sanctity of the peasant's places of Festival descended intact to both the Chieftain and his Town. . . . The seigniorial town is the successor of the Holy Place. . . . The noble's town is holy; it contains a market, an altar to the Soil, a temple of Ancestors."[19]

In the same way as in Mesopotamia, Rome, and Greece, the Chinese city, like the Japanese city that followed its model,[20] asserts itself as the monumental construction of a built environment. It represents the human world that reflects and reproduces the divine world; a holy space as opposed to ordinary space; it repeats the order of the world; or even better, guarantees that order by expressing it. From one culture to another there are profound differences in the type of monumentality in terms of style, techniques, and even mass. However, in every case, the architectonic project becomes the privileged expression that articulates human creation by replicating the creation of the divine world or in certain traditions, the creator's.

---

18. Ibid., p. 176.
19. Ibid., p. 176–77.
20. On this point see the work of N. Fiévé, *L'architecture et la ville du Japon ancien. Espace architectural de la ville de Kyôto et des résidences shôgunales aux 14e et 15e siècles* (Paris: Maisonneuve et Larousse, 1996). Note that in the case of the creation and construction of Kyoto the ramparts are absent, not so much because there are no military risks, but because the city was deemed sufficiently protected by the different religious sites surrounding it, in particular the large temples in the Northeast—a symbolically dangerous space. For a more general and thorough approach to the Japanese city, see A. Berque, *Japan: Cities and Social Bonds* (Yelvertoft Manor, Northamptonshire: Pilkington Press, 1997).

However, we must counterbalance foundational descriptions with empirical realities. The foundation and construction described in Chinese or Roman canonical texts express an ideal. Most often, the creation of cities is linked to the conquest of new territories or to reconstruction after some sort of destruction, (due to natural catastrophes, fires, or military action), or to frequent transfers of the capital as in China or Japan. Furthermore, particularities of the terrain or limited resources may lead to quite a few compromises with the rites. Canonical texts need to be put in perspective; nevertheless they allow us to envision the model that is clearly at the core of the culture in question. Moreover, it is important not to consider foundational rites as universal. Many very ritualistic civilizations do not practice them. This is the case of ancient India, which could be explained by the central importance of sacrifice which takes on all foundational functions, for both the universe and the city (as noted by C. Malamud.[21])

The Case of Greece. The radical change in the urban model in ancient Greece around the eighth century BC is most instructive. During the so-called archaic period, the city mostly consists of an agglomeration around the citadel, which is called the *astu*. The name then denotes the entire locality. We are dealing here with a highly hierarchical society, ruled by the *wanax* (later the *basileus*) surrounded by warriors on horseback that form the nobility. The rest of the city comprises artisans, among whom the blacksmiths have a privileged position, and merchants. Peasants farm the surrounding countryside and find refuge in the *astu* when attacked by foreigners.

This arrangement is profoundly modified when, as a result of various crises, military organization undergoes a transformation: every man who can carry arms becomes a soldier and the cavalry is replaced by infantry. The group of warriors invents a type of original democracy: forming a circle, they outline a neutral space in the

21. C. Malamoud, "Sans lieu ni date. Note sur l'absence de fondation dans l'Inde védique," in M. Detienne [ed.], *Tracés de fondation* (Paris-Louvain: Peeters, 1990), p. 183–191.

middle—the *meson*—where the booty is divided by lot. The circle becomes an assembly of debate and decision-making. Everyone must address the others from the common space of the *meson*.[22] This becomes the new model of the city—*polis*—where all citizens are recognized as peers—*homoioi*—and equals—*isoi*. Now the city itself is considered the *meson*, the public space where the public square—*agora*—is defined and where temples are built, as well as the theatres, the stadium, and especially the common hearth—the *Hestia koina*. One detail here is remarkable: the Greek city does not include palaces for rulers: "Archaic and classical Greece was a world without palaces or private mansions."[23] The only public monuments are religious or linked to assembly spaces—the *ecclesia*—or to cultural and sports activities such as the theatre or athletics. Moreover, temples are for public and civil cults, without secrets or priests with privileged knowledge. At the same time, the law is conceived as that which by definition should be known by all and promulgated in writing to ensure that its terms are unambiguous.

J. P. Vernant and P. Vidal-Naquet have shown how this democratic organization corresponded to a slew of geometric and cosmological representations.[24] Ideally the city is a circle where every point is at an equal distance from the center. Similarly the world is spherical and ruled by the laws of proportion and harmony, "imitated" by the city's design. This is why the same vocabulary with terms such as homology and isonomy can be found in geometry, cosmology, and politics. Architecture and the construction of cities thus belong simultaneously to those three fields. We find a good example in the work of Hippodamus of Miletus (fifth century)

---

22. This change is described by M. Detienne in *The Masters of Truth in Archaic Greece* (New York: Zone Books, 1996), p. 89 ff.

23. M. Finley, *Early Greece: The Bronze and Archaic Ages* (London: Chatto & Windus, 1970), p. 144.

24. J. P. Vernant, "Space and Political Organization in Ancient Greece," *Myth and Thought among the Greeks* (London and Boston: Routledge and Kegan Paul, 1983), p. 212–34; "Geometry and Spherical Astronomy in the First Greek Cosmology," ibid., p. 176–89. P. Lévêque and P. Vidal-Naquet, *Cleisthenes the Athenian: An Essay on the Representation of Space and Time in Greek Political Thought from the End of the Sixth Century to the Death of Plato* (Atlantic Highlands, NJ: Humanities Press, 1996).

who drew the plans for Rhodes and the city of Piraeus. In both cases he used a checkerboard blueprint, leaving a space for the agora in the center. The Greeks thought of Hippodamus as a great architect or even what we would call today an urban planner, not only because of his mastery of the art of building, but also because of his status as a "meteorologist," a scholar of celestial phenomena and the movement of the planets. Knowing the geometry of the divine world enabled him to build dwellings for humans and also entitled him to give them laws. Indeed Aristotle attributes to him a constitutional project for Miletus. Astronomer, urban planner, philosopher, Hippodamus concentrates in himself three elements linked to the very idea of the city: the sky, the political order, and thought. The space of the city reproduces the cosmic order. As a community (political community: *politeia*), the city is to humans as the world is to the human species as a whole. The construction of the city as monument is essential to the establishment of the city as a community. M. Finley sums it up as follows: "The aesthetic-architectural definition of the city was shorthand for a social and political definition: a genuine 'city' was a cultural and political center . . . a place where the well-born and educated people could live a civilized existence."[25]

THE PROBLEM OF THE CITY'S ENCLOSURE AND IMAGE

In every known civilization, ancient cities are circumscribed by a carefully delineated perimeter. The fact that this perimeter most often took the form of a rampart and had an obvious defensive military function should not mislead us. All the archaeological evidence confirms that the primary function of the rampart was religious.

For example, Roman tradition called for the ritual digging of the ditch that would become the perimeter of the city. Just such a

25. M. Finley, *The Ancient Economy* (Berkeley: University of California Press, [1973] 1999), p. 124.

ditch was dug by Romulus with a plow, according to legend. We also know that Remus was put to death for crossing it as a provocation. This ditch divides the earth in two; the outside is left to the nether gods, while the inside edge, prefiguring the rampart, delineates the sacred space of the City, of the celestial gods to be welcomed and honored by humans. The rampart does have a defensive function, but from a religious point of view, it is above all a defense against exterior and foreign space. The enemy who might attack the city is only the extreme figure of menacing exteriority. The agricultural space surrounding the city has status only as it is derived from the city that encompasses and sanctifies it in its orbit. Then, as in Rome, the original figure of the citizen-farmer emerges. Beyond that space, the wild, uncultivated areas are the beginning of a dangerous space belonging to a universe which is neither mapped nor sanctified.

Christianity, an urban religion, emulates the same distrust of rural space, the *pagus*. The pagan is the man of the fields, the *paganus*, who is beyond the spiritual space of the city and who can still be seduced by the obscure forces of the earth with its rituals and secrets. This is where, unlike in the city, polytheism persists and resists, where magical practices crop up, and where sorcerers and demons hide.[26] There are holy cities, says Duby,[27] while there is no holy countryside; at most, there are places of pilgrimage destined to sanctify the country. More generally, we see in the Middle Ages the opposition between *the bourgeois* and the *villain*; the *urbanity* of the former and the *rusticity* of the latter. The semantic charge of these terms clearly shows the hierarchy of values of the ancient and medieval world in the differentiation between the city and the countryside.

---

26. C. Ginzburg shows its long tradition in *The Night Battles: Witchcraft and Agrarian Cults in the Sixteenth and Seventeenth Centuries* (Baltimore: Johns Hopkins University Press, 1983).

27. G. Duby, *Histoire de la France urbaine* (Paris: [Ed.] Seuil, 1979), preface. Confirmed by Jacques Le Goff: "Primarily an urban religion, Christianity maintained urban continuity in Europe." *Medieval Civilization* (Oxford: B. Blackwell, 1988), p. 57.

The city marks the separation of the inside and the outside. The urban enclosure probably contributes most to making the city into a monument. It shapes the city as an individualized entity, a visible totality, and an architectural body with recognizable characteristics. However, the external unity of the city points to a more profound unity because an architectural monument is not simply a building. It is a *tectonia*, a construction that concerns the *arche*—the origin—and contains its own origin since the law of harmony gives it internal coherence. In that sense, it is an *analogon* of the world. According to Vitruvius, the different principles underlying architectural operations (*distributio*, *ordinatio*, *dispositio*, *eurythmia*)[28] all contribute to the required harmony, both of the building itself and with the surrounding site and other buildings. The foundation is the *proportio* obtained with the help of a measuring unit—the *module*—which is exactly half the diameter of the type of column chosen for the intended building. We are indeed in a universe of analogies, here the analogy of set correspondences between the natural world and the built environment. In the latter there must be an equilibrium between the different components which supposedly provides its grandeur and beauty. This makes it easier to understand the relationship of mutual expression between the city and the world, analogous to the one between the temple and the universe, as noted by a Christian writer of the seventh century: "It is such a remarkable thing that with its small size the temple should be similar to the vast universe, not so much in terms of its dimensions but as a model. Look at its high dome: it is comparable to the heavens and similar to a helmet, as its upper part rests on the lower part. Its vast, splendid arches represent the four sides of the world. On those, in those and through those, the entire roof is linked to the arches."[29]

Like the city, the temple is an *analogon* of the world. We are dealing here with interlocking models in the play of set propor-

28. Vitruvius: *The Ten Books on Architecture*, Herbert Langford Warren (Illustrator), Morris Hickey Morgan (Translator) (New York: Dover Publications, 1960).
29. Maximus the Confessor, 580–662, cited in D. Payot, *Le Philosophe et l'architecture*, p. 68.

tions. This idea of the city informs our whole culture to such an extent that the design of the enclosure still has symbolic value even though the walls disappeared with the urban expansion of the industrial age. On maps and even in administrative matters, there remains a difference between intra- and extra-muros. However, as the memory of the enclosure fades away, it foreshadows the end of the old model of the city and the transition to a novel situation. This is the beginning of the farewell to "Europe and its old parapets" (Rimbaud). On the bases of the razed ramparts the embankments of the large boulevards open up, wide canals through which the crowds of the Metropolis flow.

This does not mean that there is a continuous history extending from the monumental model of the ancient city to its apparent ending during the Enlightenment. Far from it. This is not the place to provide details, except to recall two major and greatly contrasting moments of its evolution. First, we have the medieval city, followed by the city arising out of Renaissance culture, which continues in the city of the seventeenth and eighteenth centuries.

It needs to be stressed how much the medieval city transforms or distorts the ancient model and ends up with a very different one. The perimeter is often reduced, sometimes by half or three-quarters, not as a result of planning decisions but because of the collapse of production and exchange that followed the so-called "barbarian" invasions and the fall of the Western Roman Empire. The city shrinks, blurring the initial grid pattern by accumulating buildings that encroach on the ancient monuments (circuses, theatres, and stadiums) and invade squares and avenues. Hints of the orthogonal design barely survive in many cities throughout Roman Europe. From then on, the typical city seems to withdraw in circles around the cathedral and the bishop's palace. It hunkers down in a reduced circumference with narrow interlacing streets and dense housing. It reconstructs a different model almost instinctively, that of the camps of Nordic tribes that for almost half a millennium had penetrated western and southern Europe. The city seems to return to the model of an archaic fortress, protected by ramparts but without the openness and sense of perspective characteristic of Greek and Roman cities. Moreover, it lost much of the sense and art of hygiene of the populations that these cities had previously developed. Is this pure regression, as has been claimed? This is questionable when we know that in this type of city Romanesque and Gothic art flourished and communal and civil freedoms were established.[30] This is where the artisan and merchant class freed itself from feudal power, where the most remarkable institutions of learning ever seen were created, namely universities. New logical and metaphysical argumentations were formulated and a novel relationship between reason and faith was defined. Primarily the product of urban life and culture,[31] medieval rationality was underrated for a long time, submerged as it was by the powerful movement of the Renaissance. However, what can we finally say about the city itself as monument and social space?

The indifference of medieval cities to perspective effects in the layout of streets, squares, and buildings tells us clearly we are no longer in the visual and scenic culture of Antiquity. What is newly invented reveals another resource, which we could call a culture of *vicinality*, on which we still draw today. Our gaze is not captured by long lines or grandiose openings but it moves gradually among dense habitats, according to the mostly irregular layout of streets. Often those run surprisingly into little squares that are usually accessible around corners but never have a central perspective. Most importantly, the medieval city experiments with the constant proximity of neighbors to the point where it develops a community life that seems to recover village life without losing the virtues of a cosmopolitan site and its freedom for the independent worker (that famous "liberating air"). The medieval city spontaneously developed what we call *livability*, a space where life is good and where a familiar feeling associates bodies and spaces. It is a space

30. On its symbolic and theological wealth, see E. Panofsky, *Gothic Architecture and Scholasticism* (Latrobe, PA: Archabbey Press, 1951).
31. See Jacques Le Goff, *Intellectuals in the Middle Ages* (Cambridge, MA: Blackwell, 1993).

where it is pleasant to walk and to meet.[32] Does it need to be said that today this should be the priority for all city planners and architects?

It is easy to understand why religious buildings acquire great importance in such relatively closed spaces of local connections. Distance and infinity are best seen when cathedrals loom up suddenly among low houses. From the narrow square in front of the cathedral, the gaze is invited to move up along the façade to the spires; in the nave, following the lines of the pillars it can soar toward the arches to lose itself in the light of the stained glass windows. The contrast between living space and church space, which is probably comparable to the one between the medina and the mosque in the Maghreb and the Middle East, sums up the vicinality/verticality relations defining the urban atmosphere of the medieval European world. The atmosphere is dominated by the intensity of the social bond and religious faith and marked by tension between physical realism and ascetic spirituality. One could show that the vicinal and vertical configuration continued for several centuries in the neighborhoods of large metropolises, albeit with different adaptations. It remains a model that still haunts the dreams of city dwellers and contemporary city planners. The equivalent can be found in the coupling of a modest shop and a skyscraper in Manhattan or in the maze of small houses behind a glass and steel curtain of skyscrapers alongside the large avenues of Tokyo.

We still have questions about the monumentality of medieval cities. Did they not lose that essential characteristic by compressing and blurring the alignment of streets and sometimes dismantling temples and monuments? That may be so. However, the medieval city did something quite extraordinary: while its parts are less monumental, it becomes more so as a whole. The ramparts that clearly indicate its limits, the tightly massed houses dominat-

32. There is no question here of idealizing medieval dwellings that often suffered from two major defects, poor hygienic conditions and crowding in spaces with little sunlight.

ed by the church towers and roofs or the high silhouettes of belfries and castles, make it look like a large, densely packed body. Even in the case of a modest city, such an organic, compact whole presents the very image of the City more clearly than the large ancient cities. It shows the uniqueness of the *locus urbanus* abundant in medieval iconography, whose ideal survives until the Enlightenment.

Meanwhile however, the experience and representation of space underwent major changes, which had a direct impact on architecture and urbanism until the birth of Modernism. These changes, which mark a profound break with the medieval world, take place in two phases. The first is technical and geometric; it starts in the fifteenth century in Italy with the *Quattrocento,* and concerns the theory and practice of *perspective*. The second could be called cosmological: it starts with the Copernican heliocentric theory (1530/1543) and ends up in Galileo's physics (1604). How does this concern the evolution of the city, at least in Europe, and its destiny as monument? Briefly stated, the first phase during the Quattrocento consists of a revolution in architecture and the visual arts that immediately affects the construction or transformation of cities. The second phase deals with a conception of matter and space that is identified with rationality itself and which tends to make the city its most global and complete cultural expression.

Neither Antiquity nor the Middle Ages were unaware of perspective; far from it. At most, they may not have solved some problems or dealt with them differently, as we know better since the late nineteenth century. However, in terms of the question that interests us, it remains true that the Middle Ages no longer considered perspective as a necessary requirement for urban planning.[33] One could say that in medieval cities streets dodge in and

---

33. Even if this concept appears very late in our languages (beginning of the twentieth century), its reality is as ancient as the construction and management of cities and their territories. See F. Choay, *L'urbanisme, utopies et réalités: Une anthologie* (Paris: Seuil [1965] 1979). The concept can also be attributed to Ildefons Cerdá, who was responsible for the renovation of Barcelona (1856), and who launched the term "urbanization" in his *Teoría general de la urbanización* (Madrid, 1868).

out of housing volumes and pierce random openings while squares appear to exist to relieve pressure by pushing back façades. On the contrary, the ancient city organized neighborhood volumes to align with streets, avenues, and the perimeter of squares that were first traced and allotted. Voids determined solids. Renaissance urban planning literally returns to this vision. It is complemented by new knowledge about the laws of perspective that are familiar to painters, architects, or the numerous people who deal with mathematics. Without entering into the complex history and detailed analyses of this knowledge,[34] we would still like to indicate some of the consequences of this *scienza nuova* for the creation of urban space.

First, there is a kind of fusion between space as treated by painters and the space developed by architects, especially since many of them exercised both professions. This interpenetration has important consequences for the concept of the city. It means that space is first conceived as an *image*. Remember that the initial technical problem posed by perspective is to represent three-dimensional space on a flat plane, *from the point of view of the spectator*. Paradoxically then, an *artifice* is needed to reestablish the *truth*, at least for the perceiving subject. And so the city becomes something *to look at*. The entire city becomes a setting for the stage. This is clearly seen in the theatre decors that were one of the most common representations of perspective. Space as a stage observed by spectators triumphs everywhere over the tactile, discontinuous, mobile space experienced by inhabitants and pedestrians. Still more radically, space is only valued as geometric space and acknowledged as "true" when its forms are most abstract. While actual neighborhoods are not necessarily affected—life always resists—the vicinality that accounted for the originality of medieval cities tends to be devalued in discursive and graphic representations. The geometric training of the gaze imposes its

34. On this vast subject see, for example, E. Panofsky, *Perspective as Symbolic Form*, MIT Press, 1991; Hubert Damisch, *The Origin of Perspective*, Cambridge, MA, MIT Press, 1994.

priorities on bodies whose living spaces are multiple and irregular. Seeing gets the better of living. Control over distance excludes the random proximity of encounters. Not only do perspective values constitute the position of the perceiving subject, the sovereign and legislative "I" that Modernity brings about, but they also shape the figure of the Prince in a public space henceforth understood as the stage *of his authority*. This authority no longer derives from his dynastic or religious status, but it is constructed, conquered, and imposed by a privileged position on the stage where he can see and be seen from an advantageous angle. This could be called the *technical turning point* of the Machiavellian moment.[35]

It is also the turning point in the conception of public space. No longer the space of the Ancients designed for citizen debates according to set rules—the exchange of words—it is a space constructed to ensure that the position of optimum visibility coincides with domination, that is, the construction of the gaze. These visual or even theatrical values underlie the construction or renovation of urban space, for instance, the large clearings initiated by the sixteenth-century popes who remodeled medieval Rome into a Renaissance city. Of course, the cities of the Italian *Rinascimento* are not simply abstract visual machines; renovations were negotiated with the medieval legacy which remained powerful. It may even be that this mixture of genres accounts for their exceptional charm.[36]

However, at a deeper level something began to change. One of the essential aspects of the theory of perspective concerns the horizon where the lines of sight converge asymptotically at the vanishing point. That point is like an analogue of infinity. The terrestrial space, whose finitude and fragility the Middle Ages constantly recalled when faced with the perfection of the celestial

35. See M. Hénaff, "The Stage of Power," in *Substance* 80, "Politics on Stage: The Machiavellian Moment and the Birth of Perspective," p. 16–21. See also Gérald Sfez, Machiavel, *La politique du moindre mal* (Paris: PUF, 1999), "La visibilité du Prince," p. 115 ff.

36. This fascinated C. Sitte, described by him in his *City Planning According to Artistic Principles* (London: Phaidon Press, 1965).

world, now seems to find its own rigorous infinity. The new cosmology elaborated by Copernicus and Galileo confirms and develops this certainty, bequeathing a spatial model of rationality that architecture and urban planning have never ceased to adopt and promote ever since. We will outline a few of its essential points here.

What has been called the Copernican revolution not only demonstrated that the center of our universe was defined by the position of the sun and not the earth but, more profoundly in the long term, it meant that we knew nothing of the center of other universes in our galaxy. The newly acknowledged centrality of the sun is not just a simple shift that is easy to absorb intellectually and emotionally. It becomes a unique case, which leaves us in ignorance of other systems. The cosmos is likely to have multiple centers. The ancient sphere has exploded into myriad worlds, a vision whose harshness, rigor, and pathos will later be pointed out by Pascal: harshness, because the center is everywhere and the circumference nowhere; pathos, because "The silence of infinite spaces frightens me." A new certitude emerges: the physical universe is infinite. This was in any case Giordano Bruno's conclusion; it was partly why he was burned at the stake.

Let us return for a moment to 1543, the year of the publication of Copernicus' theses. Historians of science have noted that in the same year another major scientific work was published, Vesalius' treatise of anatomy, which revolutionized the observation and description of the human body. Not long before that, in 1538, Mercator had published the first scientific maps of the earth, according to the cylindrical projection method named after him. A revised edition appeared in 1569. Thus, in a few years a remarkable upheaval took place in the representation of space and the body. There emerged a kind of correspondence between the cosmos, the earth, and the human body that is *no longer symbolic* but proceeds from a *rigorous assessment.* These relations were always at the heart of architectural thought; they continued to be but with profound changes in their program.

Let us now turn to the second decisive phase of this vast mutation, that is, Galileo's physics. Beyond confirming Copernicus' cosmology, it unifies the understanding of the universe by assuming it is written in mathematical language. All bodies can be understood in terms of extension and movement, including living bodies. Descartes draws all the philosophical conclusions and contributes to the formation of a paradigm going far beyond the geometry of perspective. From now on, the universe is assumed to have substantial homogeneity and its forms to have a calculable regularity. It seems obvious now that humans are responsible for bringing their productions into conformity with the truth of the physical world, especially in what most immediately expresses this aspiration for unity, the built environment. Architecture can do no less than to follow in the footsteps of physics, starting with its largest construction site, the city. Not surprisingly then, at the very beginning of his *Discourse on Method*, Descartes explains his project of completely rethinking the foundations of knowledge by comparing it to the task of an architect who finds it much more efficient and rational to start from scratch when faced with the necessity of renovating an ancient city. Nothing is comparable, says Descartes, to the sites drawn on a flat surface by an engineer following his imagination. This ideal of a tabula rasa will haunt architectural thinking for centuries, as for instance at the École des Ponts et Chaussées (School of Bridges and Roads—founded in 1747) and before that, among military engineers. The latter designed fortresses buried in leveled terrains, with overhanging ramparts at sharp angles to divide the attackers from above. The architecture that had absorbed the lessons of the Italian Renaissance and recovered the ancient ideals is faced with a new challenge. It must integrate the requirements of accuracy and calculations of Galileo's physics into this legacy still saturated with symbolic and numerological considerations. As seen in the quarrel between the Ancients and the Moderns,[37] this tension has often been fraught

37. On this point, see A. Pérez-Gómez, *Architecture and the Crisis of Modern Science* (Cambridge, MA: MIT Press, 1983).

with conflict. Nevertheless, it has produced some balanced monumental works: in Paris, for instance, the Invalides and the École Militaire and their esplanades, the Place de la Concorde and its palaces, and the Avenue des Champs-Élysées. However, toward the end of the eighteenth century, the requirement of mathematical rationality would lead to attempts at close and often narrow conformity between architectural forms and social needs, as we will see later.

Let us summarize our thought on the city as monument. As long as there have been cities, and for whatever the particular reasons for their emergence within a given civilization, one thing seems certain when we consider the symbolic organization of urban space and the institutional forms they engender: *the city is built and organized to take shape as a world in itself.* What does this mean? The city is not simply a palace where the prince and his court live, nor a monastery with a limited number of men or women, nor a fortress where soldiers are on watch, all instances of monumental construction. It is the place for *all* the members of the community to live; the city also absorbs and organizes the surrounding space around itself. This is why all the world's elements must be concentrated in the city: humans, deities, the natural order, and the world of artifacts. It articulates the earth and the sky heavens and presents itself as the summary of the universe. In the city's perimeter, public and private buildings *together* form a monumental totality where ideas, beliefs, and shared sensibilities are exchanged. Through its constructed form, the city attempts to realize a spiritual unity and organic whole. The city is the architectural body of this desire or intention, which confers exceptional dignity to architecture. The city is not only a place with monuments: *it is itself a monument par excellence.* It is the creation that includes all creations, which welcomes projects and their implementation because it is the place that ultimately expresses the relation between origin—*arche*—and what is built—*tekton*.[38]

The Greek and Roman legacy may have echoes and similarities in many other cultures but in the case of the West, it intersects remarkably with the legacy of the biblical tradition. In that respect, the most illustrious text remains the *Apocalypse* of John which set for a long time the image of the celestial City:

> And I saw a new heaven and a new earth . . . And I John saw the holy city, new Jerusalem, coming down from God out of heaven, prepared as a bride adorned for her husband/ / And I heard a great voice out of heaven saying, Behold, the tabernacle of God is with men, and he will dwell with them, and they shall be his people, and God himself shall be with them, and be their God / / And there came unto me one of the seven angels And he carried me away in the spirit to a great and high mountain, and shewed me that great city, the holy Jerusalem, descending out of heaven from God/ / Having the glory of God: and her light was like unto a stone most precious And the wall of the city had twelve foundations . . . And he that talked with me had a golden reed to measure the city, and the gates thereof, and the wall thereof.[39]

The celebration of the city, its promotion to a divine and holy site uniting God and humans, had rarely attained such metaphorical range. There are doors carved in giant pearls guarded by angels, ramparts made of precious stones, and pavements of pure gold. The exaggerated description is possible only because of a shared certainty: at the end of time, the forever lost Garden of Eden will be replaced, not by another garden, but by the holy city. The city as such will be the residence of humans on earth and the site to receive the divinity. In Christian thought, the human city becomes

38. This is much emphasized by D. Payot in *Le Philosophe et l'architecte*, p. 68—from this perspective, the early modern city best exemplifies the model of analogic thought, outlined by P. Descola next to the other three (animic, totemic, and naturalist) in his important work *Beyond Nature and Culture* (Chicago: University of Chicago Press, 2013).

39. John, *Apocalypse*, Rev. 21:1–16 (Bible, King James Version).

the city of God. Urban space itself becomes the figure of the saved.

We can say then that from its emergence the city as monument and as microcosm reflecting the world becomes the canonical figure of the space par excellence where humans live. It expresses a totality, a symbolic unity, and often holiness. It is also the site of the community's institutional grandeur, the built space of its public existence, and hence of political sovereignty. Finally it is the privileged space of knowledge and the arts. At the beginning of *Phaedrus*, Socrates affirms: "Neither the countryside nor the trees teach me anything, whereas men in the city do (*Phaedrus*, 230 d–e). The word chosen here for "city" is not *polis* (the community ruled by law) but *astu*, which designates the built totality, the residents' material physical living space. For Socrates, the city is already *the site for thinkers* par excellence.[40] It is the place for humans to mix with the gods or imitate them since it inseparably integrates monumental architecture and public life ruled by law. This may not be the case outside the West, in China or India for instance.

As prints and texts show, the city's form seems to remain intact until the end of the eighteenth century. Curled up in its enclosure and rooted in the landscape, the city is most often a harmonious monument where humans live and act. At most, people complain about traffic congestion in the narrow alleys or about poor sanitary conditions. Even Rousseau, who detests Paris because of its overpopulation and hypocritical social relations,[41] salutes its architectural splendor. The urban unity and totality as monument, as the *compendium mundi*, the summary of the world, appears most clearly in a genre painting of the cityscape. Very popular in Europe since the Renaissance,[42] it shows the city's monumental scen-

---

40. Cf. M. Hénaff, *Le lieu du penseur* (Paris: Cahiers de Fontenay, 1983).

41. Cf. M. Hénaff, "The Cannibalistic City: Rousseau, the Large Number and the Abuse of the Social Bond," *Substance* 63 (1992).

42. A good example is the *Liber Chronicarum* (1498) by Hartmann Schedel that offers a series of "views from above" of a number of famous cities (some being imagined).

ery seen from afar as an architectural whole. There are quite a few urban *vedute*[43] from the most well-known cities (for example, Florence, Siena, Toledo, Paris, Salzburg, Heidelberg, Dresden, Prague, Krakow, Moscow, Delft, and Ragusa) to much smaller ones. Each city's uniqueness appears in its profile drawn by the lines of the ramparts, churches or palaces emerging from the mass of roofs and houses, the "panorama," "skyline," or "*Stadtbild*," more a portrait than a landscape. It is as if the monumentality glimpsed from afar by the traveler eventually gave the city a unique face corresponding to the originality of its name. The Flemish painter Joachim Patinir develops an even more significant genre around 1550, continued by his student Cornelis Massys and sanctioned by Peter Breughel (1569–1625). The genre depicts in an imaginary panorama called a "universal landscape" the elements of a world often focused on a city that is surrounded by the sea, a forest, a lake, the countryside, a river, or a desert.[44] At the heart of Renaissance humanism, the city represents the totality of human achievement and appears less as something that springs up in the world than as the site that gathers the world in itself. The city is the world.

43. On the city as monument and landscape and its organic unity, there are a number of important works, particularly of scholars of German or Nordic origin such as Wolfgang Braunfels, *Mittelalteriche Stadtbaukunst in der Toskana* (Berlin: G. Mann, 1953 and *Abenländische Stadtbaukunst* (Köln: Dumont, 1976); Christian Norberg-Schulz, *Genius Loci: Towards a Phenomenology of Architecture* (London: Academy, 1979), and *Architecture: Meaning and Place* (New York: Electa/Rizzoli, 1986); Tilo Schabert, *Die Architektur der Welt: Eine kosmologische Lektüre architektonischer Formen* (München: Fink, 1997).

44. Cf. Hans Seidlmayr, "La ville comme oeuvre d'art" in *La Revue du MAUSS* 14 (2nd sem. 2002), translation by J. Dewitte.

2

A MACHINE: THE CITY THAT ORGANIZES, PRODUCES, AND TRANSFORMS

The holistic image of the city, defined ideally by its monumental silhouette, is well established, as is apparent in the pain felt for the destruction of its former equilibrium: "Ancient Paris is no more. The form of a city changes faster, alas, than the heart of a mortal." This is Baudelaire's well-known complaint ("The Swan," in *Flowers of Evil*) when he sees Paris disemboweled by Haussmann's renovation. The industrial revolution led to the "explosion" (to use a questionable metaphor) of the European traditional city; however it was also a global phenomenon. This revolution is often described, by Mumford or Toynbee for instance, as a kind of history external to the city itself. It is seen as an uncontrollable evil force that suddenly appeared from another region of our civilization, overwhelming our cities and ruining forever the organic unity of the quiet enclosure that mirrored the heavens and summarized the world, in which nature's cipher and the rational order were inscribed.

We should probably reject this legend and hypothesize that the *industrial revolution was the direct consequence of the city's very success* as a monument. In other words, it followed from the inherent logic of the urban phenomenon since its beginnings. In all sorts of civilizations, the city implied something of a very different order that went beyond mere monumental splendor. The essential element of the urban phenomenon is the fact that the city is a machine and even a *mega-machine*, to use Mumford's phrase. We will have to elaborate the implications of his brilliant intuition. However, this machine can only operate and produce where an administrative framework guarantees the organization of the population, and provides services and a stable functioning order. Only under such socio-technical conditions can the organizational system become an engine of economic growth. Let us now consider these three aspects.

THE TECHNICAL SYSTEM AS SOCIAL MACHINE

The mega-machine must be understood as the social system that organizes *labor*, especially in coordinating large building projects. Many of Mumford's analyses are contested today, but on this point he formulated the problem correctly:

> The many diverse elements of the community hitherto scattered over a great valley system and occasionally into regions far beyond, were mobilized and packed together under pressure, behind the massive walls of the city. Even the gigantic forces of nature were brought under conscious human direction; tens of thousands of men moved into action as one machine under centralized command, building irrigation ditches, canals, urban mounds, ziggurats, temples, palaces, pyramids, on a scale hitherto inconceivable. As an immediate outcome of the new power mythology, the machine itself had been invented: long invisible to archeologists, because the substance of which it was composed—human bodies—had been dismantled and decomposed. The city was the container that brought about this implosion, and through its very form held together

> the new forces, intensified their internal reaction, and raised the whole level of achievement.[1]

It is important to focus on the city as a unique phenomenon that concentrates a laboring population. The opposite of ephemeral nomadic gatherings, the city is unique in its spatial stability and temporal continuity; it is unique in terms of size compared to village settlements. Finally, it is unique in its organizational structure, which no longer rests on kinship systems or personal allegiance. One could say that the city combines the three legacies while extracting one element that reduces their impact.

Empirically the city is the continuous concentration of a large number of individuals in the built environment of a restricted area. This alone constitutes a totally new technical and social phenomenon:

- *technical* first of all, because to build presupposes the availability of means, such as large quantities of materials, stones, bricks, wood, and a diversity of specific tools, as well as the know-how of builders and artisans and a large workforce. In terms of scale and specialization, those are all things that represent a break with the rural way of life.
- *social*, because given its concentrated population, the city can only subsist by imposing a division of labor that includes the systematic organization of complementary tasks. This is why the city no longer values kinship relations but prefers functions and professions. Even when the values of ethnic traditions survive, as we see in Greece and Rome, and in eighteenth-century European cities and later, professional activities increasingly define hierarchy and status. We see this in the medieval city in the West, where a hierarchy often emerges elevating decision making over implementation. What made this possible is a profound change in the very concept of *work*, which needed to be rescued from the con-

1. Lewis Mumford, *The City in History* (New York: Harcourt, Brace & World, Inc., 1961), p. 34.

> tempt in which it was held by the classical world. With the emergence of Christian teachings, work became the natural condition of sinful humanity and a legitimate activity associated with the search for salvation.[2]

The city effects a huge transformation in the organization of labor. On one hand work no longer functions according to seasonal rhythms; the city retreats from nature, specifically by defining the nonurban as "nature." On the other hand it detaches individuals from prescriptions linked to age, lineage, and gender difference in order to consider them in terms of their technical or administrative competency. This is definitely a retreat from tradition. Moreover, for large construction projects, the city treats individuals as simple reservoirs of forces to be organized and assigned tasks defined by others who have specialized knowledge like geometers, architects, or engineers, or have power like civil servants, guards, and inspectors. Very early on the city concentrates forces as pure forces. This means that well before the appearance of the factory that Marx discusses, urban organization extracts human forces as quantities, as purely mechanical energy. It can then be used and coordinated with other forces to obtain more power for the construction of collective structures such as walls, pyramids, fortresses, palaces, temples, or canals or any other type of collective project.

We are dealing here with a techno-social machine because the city is the first type of organization that assembles, distributes, and coordinates living organisms to create collective technological projects. As Mumford correctly noted, this social machine is a *complex machine*; archaeology cannot say much about it since it only existed as the organization of bodies and disappeared with them as well. Only tools subsist here and there and they cannot give us much information about the collective distribution of tasks. This mega-machine itself constituted the privileged center of the major

---

2. See J. Le Goff, *Time, Work and Culture in the Middle Ages* (Chicago: University of Chicago Press, 1980).

technological change from which would proceed all other inventions and their widespread usage. The city anticipates the transfer of the "machine" from bodies to manufactured instruments that are linked together in more and more autonomous wholes with their own logic.

Village civilizations could not even imagine such a transformation. While the village persists as the model of a living organism, the city begins to look more like an automaton. This is why, again according to Mumford, the city appeared as "a storehouse, a conservator and accumulator. It was by its command of these functions that the city served its ultimate function, that of transformer."[3] Well before the emergence of industrial society, the city started to break up traditional bonds and invent an order linked to individual capacities and initiatives. It constitutes the first mechanism exploiting the biotope on a gigantic scale with a rational organization.

The city is a *world* in a different sense from the tradition that saw it as the image of heaven and the microcosmic double of the universe. The city creates a world by changing the world and multiplying its artifacts. Moreover, it attracts every manifestation of the culture in which it emerges. No longer simply the analogical double of the cosmos, *it becomes the world it makes.* As a megamachine, the city breaks away from the heaven whose reflection it claimed to be and ceaselessly engenders its own dynamics producing forms and goods. The city begins to face nature as its other, transforming it into matter to be measured, molded, and exploited. The city is the techno-social crucible that changes the world. It is the environment where all sorts of techniques multiply, generated by the concentration of groups of artisans and their mutual interactions. We are dealing here with a threshold effect that could explain how the city in every civilization became the intense center of *innovation.* For centuries, the city remains the space for intellectual and technological invention. However, this cannot be explained solely by the concentration of know-how and tech-

3. Mumford, *City in History*, p. 97.

niques. It happened also because the city's social milieu enables city dwellers to break away from old allegiances and to meet visitors from elsewhere such as merchants, pilgrims, seasonal workers, immigrants, or refugees. They mix with different ethnic groups from neighboring or distant regions. The constant mixing and cultural blending of the city offer the pleasures of novelty and surprise, as well as far more freedom than villages and clans. This is already the "freedom of the Moderns." Innovation prospers in this framework of multiple experiences and unexpected encounters, in a milieu that gratifies and promotes *free curiosity* instead of being suspicious of it.[4] The city is seductive because the curiosity of malleable minds cannot be separated from a taste for freedom inspired by a sense of time as unlimited promise.

As a productive machine, the city simultaneously mobilizes artisanal know-how and calls for formal knowledge. It becomes the crucible where all knowledge converges, schools are born and prosper, where engineers and artists settle, and culture itself becomes dense. Urban society is incessantly carried and driven by this dynamics of creation and change. Nevertheless, monuments continue to occupy the epic scene and to shape the city's image. They continue to forcefully call for signs from above since their function is still to celebrate the gods and to legitimate the prince, to honor the great and to legitimize institutions.

However, monuments include another dimension which tends to be ignored, that is, they are produced by *collective labor* without which they would not exist. Monuments testify to this labor even when the work involved is forgotten and the created forms relegate the *technē* that shaped it to the background. Almost behind the scenes, the city as architecture ceaselessly creates and reproduces itself as a mega-machine. As a built complex it permanently testifies to the industrious energy and productive intelligence that made it possible, revealing the ongoing process of con-

4. The history of curiosity, or thirst for knowledge or the unknown, remains to be written. Precious elements can be found in Hans Blumenberg, *The Legitimacy of the Modern Age* (Cambridge, MA: MIT Press, 1983), part III.

struction, development, and sometimes reconstruction. The built environment can only exist through the work of the social megamachine: monuments, public buildings, the layout of dwellings and their functioning within a circulation network. A monument always presupposes a machine. Each stone of a beautiful palace recalls the labor and technique that created it. The urban phenomenon has another dynamics that no longer relates to monuments but to machines. It does not signal glorious signs and works but the power of techniques, productive activities, and manufactured things, often accompanied by human effort and suffering.

However, it must be noted that artisanal techniques, and especially metallurgy, were already considerably developed in village societies. What is new in the city is their concentration and coordination for *collective projects*. Such a change could only come about because of the emergence of management technology involving group organization. The monument-machine becomes possible only through powerful organizational engineering. Let us consider this further.

THE CITY AND ADMINISTRATIVE RATIONALITY

Only in the city does there appear a hitherto nonexistent type of organization, that is, *administrative organization*. To be sure, it is linked to the appearance of what since the sixteenth century is called the state, but it emerges only in urbanized states. Centralized government existed among nomadic peoples such as Scythians and Mongolians but they did not have an administrative system proper, only hierarchies of chiefs and servants, which are very different from a civil service. The latter constitutes a quasi-autonomous body, different from a political or military hierarchy. When the Mongolians for instance settled in Chinese cities they adopted the Chinese administrative system since they did not have one of their own.

What does *administrative rationality* mean? This expression comes from Max Weber, for whom the concept of rationality, applied to social relations, is always the sign of axiological neutrality. This means the emancipation from traditional dependency or purely local rituals. There is rationality when some function can be universalized because of verifiable efficiency; this becomes the foundation of its legitimacy. We can summarize a few essential elements of Weber's analyses of the characteristics of a state bureaucracy.

First of all, the relation between a civil servant and power is not based on a personal relation whether it be charismatic, feudal, or linked to kinship, but on an abstract necessity defined by rules concerning functions to be fulfilled. The civil servant is a cog in a system that implements decisions and maintains regulations. Even if he is hired by those in power and depends on them, the civil servant has real independence to make judgments and decisions since his actions are supposed to obtain results in his area of competence, and not to express allegiance to political power. This of course is true in principle, as an ideal-type.

Secondly, civil servants are linked through hierarchical relations in the internal organization of a system that defines tasks according to different levels. Each only deals with his or her domain of competency and does not intervene in what belongs to a superior higher rank. This delimits both one's responsibilities (and privileges and risks) and rules of authority.

Thirdly, the appointment of civil servants is based on competence and merit. This competence is recognized empirically or established by selection procedures based on exams or various competitions. A bureaucracy is a management instrument founded on knowledge relative to domains of action: "The primary source of the superiority of bureaucratic administration lies in the role of the technical knowledge which, through the development of modern technology and business methods in the production of goods, has become completely indispensable."[5]

---

5. Max Weber, *The Theory of Social and Economic Organization*, ed. by Talcott Parsons (London: Free Press, Macmillan Publishing Co., 1964), p. 337.

This explains the very ancient link between administration and the mastery of writing techniques in Sumer, China, and Egypt. Weber adds:

> Bureaucratic administration means fundamentally the exercise of control on the basis of knowledge. This is the feature of it which makes it specifically rational. This consists on the one hand in technical knowledge which, by itself, is sufficient to ensure it a position of extraordinary power. But in addition to this, bureaucratic organizations, or the holders of power who make use of them, have the tendency to increase their power still further by the knowledge growing out of experience in the service.[6]

Fourth, administrative functions are full-time and must remain the sole occupations of those exercising them, which avoids conflicts of interest and possible corruption. Furthermore, one's career is generally guaranteed and promotion rests on seniority. In return, civil servants are supposed to be discreet or neutral vis-à-vis the population, as well as the authorities they serve.

Those are the essential characteristics of what Weber call an "ideal type." It goes without saying that multiple nuances can be found depending on various traditions, peoples, and situations. No matter what problems arise from dysfunction, such as corruption, nepotism, obstruction, or incompetence, Weber believes that the administrative organization of states, both ancient and contemporary, is the very condition of their existence and stability, because it proves their rationality.

Weber does not simply acknowledge that administration is an essential feature of the state but he also claims that it is essential for the functioning of the whole society, especially of economic activities: "Though by no means alone, the capitalist system has undeniably played a major role in the development of bureaucracy. Indeed, without it capitalist production could not continue . . . Its development, largely under capitalist auspices, has created an

6. Ibid., p. 339.

urgent need for stable, strict, intensive, and calculable administration."[7] This necessity is true in many domains: justice, education, health, the army, culture, transportation, industry, commerce, maintenance, and the police. However it must be noted that the emergence of bureaucracies in centralized states is inseparable from urban culture. The management of urban populations requires organizational and regulatory methods that do not seem necessary in rural populations subject to traditional dependencies primarily defined by lineage, and where the demographic level remains modest. Bureaucracy appears with the urban machine or mega-machine, and in return is essential to its proper functioning.

Regardless of state structure, as a built environment concentrating a large population, the city had to generate a variety of specifically technical services without which the population could not survive. The case of Rome is particularly instructive.[8] The prefect of the city (*praefectus urbis*) had authority over most urban services, particularly over public records, archives, finances, and education. The prefect of the city takes control over all the main public services such as the police, firefighters, food supplies, water management, public buildings, the Tiber, and the sewer system. One example are the functions of the *vicomagistri* (*masters* of the *vicus* or neighborhood) who are charged with watching over places of worship, the control of markets, the representation of the neighborhood to higher ranked administrators, and collaboration with vigils to maintain peace in the streets, day and night. Other services entrusted to the care of guardians are the upkeep of monuments and statues, organizing vigils to guard aqueducts, riverbanks, the port on the Tiber, managing the sewage system and the supply of different foodstuffs such as grains and wine.

All this should give us an idea of the complexity of the organization and management of a city as far back in time as we wish.[9] To

7. Ibid., p. 338.
8. Léo Homo, *Rome impériale et urbanisme dans l'Antiquité* (Paris: A. Michel, [1951] 1971).
9. For details about urban services in Greece, see Roland Martin, *L'urbanisme dans la Grèce Antique* (Paris: Picard, 1956).

exist, the city calls for an *administrative machine* from which it is inseparable. Weber's thesis is beyond question: the administrative organization indeed means the advent of rationality. However, he should have specified that such rationality only appears with the mega-artifact of the city and is necessary to make life possible by ordering it in a system of complex relations between humans and artifacts. Administration at first is nothing more than the social engineering that living together in the built-up space of the city requires. The city then becomes the most complete and complex technical environment ever created.

ECONOMIC DYNAMICS

Clearly, the city's eminent role in economic development is linked to *its intrinsic technical dimension*. However, we must have serious reservations about Mumford's notion of the mega-machine, which primarily concerns large building projects. Those projects are not permanent and normally lie outside the city like the pyramids in Egypt or the Great Wall in China. The urban accumulation of trade associations and artisanal production becomes a long-term reality. This permanent feature specifies the volume of production as well as the specialization and accumulation of know-how in a limited space, which produces favorable conditions for technological innovation, as we have seen.

However, the technical concentration and division of labor characteristic of urban space are not enough to explain its economic development. An element of a different order was needed, namely *exchange*. In the first Mesopotamian cities the necessity of importing materials entailed the export of goods locally made. But apart from foreign trade, which also constitutes an essential network dimension, the city can only function internally as a *market* given the differentiation and specialization of crafts. Aristotle understood this quite well. He saw the material cause of the city's existence in the community of interests, which he calls *koinonia*, and in exchange regulated by money the adequate translation of reciprocal needs, while the final cause was the formation of a political community, the *polis*.[10]

The existence of an internal and external market is certainly one of the major criteria of a city's economic dynamics. One must agree with most contemporary historians that the advent of the city is not due primarily to economic factors but rather to the introduction of a political and administrative authority.[11] Nevertheless, as soon as the city emerges, it becomes a market. Again, this is a crucial point in Weber's thesis. "In the meaning employed here the 'city' is a market place."[12] Wherever it appeared as a configuration different from the country, it was normal for the city to be both a lordly or princely residence as well as a marketplace."[13] Weber understands however that it is not enough to only consider the concentration of productive forces. For the city to go beyond purely local production and for accumulation to appear, a number of other conditions were necessary. Those were legal and political. A relatively free market in land and real estate was needed, but above all, a proper municipal legal system had to liberate the city from feudal tutelage in the justice system, administrative management, civil defense, and finally in the political organization of the community. According to Weber, "Such a law was absent from the courts of the major number of the cities of the world."[14] The originality of cities in the West consisted in developing this legal system, often at the cost of fierce battles. Since the Middle Ages, "In varying degrees the city became an autonomous and autocephalous institutional association."[15] After the urban renaissance of the twelfth century this autonomy gives rise to exceptional movements of commercial exchanges and releases a new logic of capital accumulation. On the contrary, legal autonomy is lacking particularly in the cities of the Middle East

---

10. Aristotle, *Politics*, book I, 1252a–1253a.
11. Georges Duby, *Histoire de la France urbaine* (Paris: [Ed.] Seuil, 1979); preface.
12. Max Weber, *The City* (New York: The Free Press, 1958), p. 67.
13. Ibid., p. 67.
14. Ibid., p. 106.
15. Ibid., p. 105.

and China. "However, the possession by the urbanites of a special substantive or trial law or of courts autonomously nominated by them were unknown to Asiatic cities."[16] This diagnosis was confirmed by É. Balazs in his study of Chinese cities.[17] Actually, this was already a problem between Western Europe, which very quickly adopted urban modernity, and eastern Slavic Europe and the Russian empire, where cities remained subject to feudal law. This is why the history of the rise in power of European capitalism in northern Italy, the Rhine provinces in Germany, the Paris-Lyon axis, the Netherlands, the Baltic coast, and England is foremost the history of urban dynamics. We see this demonstrated by F. Braudel[18] and later E. Wallerstein.[19]

Not surprisingly then, the industrial revolution which started in Europe in the second half of the eighteenth century and accelerated in the first half of the following century was mostly linked to urban centers. Those cities where big industries are concentrated experience unprecedented demographic growth, increasing their population three- or fourfold. Then follows the disorderly development of the housing real estate. It culminates in the sinister suburbs dominating the urban question of the nineteenth century and which still remain a major problem in contemporary metropolises. According to Sigfried Giedion:

> The industrial revolution, and the abrupt increase in production brought about during the eighteenth century by the introduction of the factory system and the machine, changed the whole appearance of the world, far more so than the revolution in France. Its effect upon thought and feeling was so profound that even today we cannot estimate how deeply it has penetrat-

16. Ibid., p. 81.
17. É. Balazs, "Les villes chinoises: Histoire des institutions administratives et judiciaires" in *La Bureaucratie céleste* (Paris: Gallimard, 1968), p. 206–29.
18. F. Braudel, *Civilization and Capitalism 15th–18th Century* (New York: Harper & Row, 1982–1984).
19. I. Wallerstein, *The Capitalist World Economy* (Cambridge: Cambridge University Press, 1979). Henri Pirenne deserves credit for having understood quite early the economic dynamic of European cities. See H. Pirenne, *Medieval Cities: Their Origins and the Revival of Trade* (Garden City, NY: Doubleday, 1956 [1925]).

> ed into man's very nature, what great changes it has made there. Certainly there is no one who has escaped these effects, for the Industrial Revolution was not a political upheaval, necessarily limited in its consequences. Rather it took possession of the whole man and of his whole world. Again, political revolutions subside, after a certain time, into a new social equilibrium, but the equilibrium that went out of human life with the coming of the Industrial Revolution has not been restored to this day. The destruction of man's inner quiet and security has remained the most conspicuous effect of the Industrial Revolution. The individual goes under before the march of production; he is devoured by it.[20]

The social revolutions and political struggles that marked the history of the whole nineteenth century were inseparable from urban upheavals. The exploitation of labor went together with the manipulation of land rent and a housing crisis, as recorded by the young Engels in one of the best analyses in his pioneering report that includes rigorous documentation and numbers, *The Situation of the Working Class in England*[21] published in Germany in 1845. He describes the intolerable poverty and unsanitary conditions of Manchester's working-class housing. Many equally alarming reports appear also in France and England at that time.

However the urban crisis is not limited to the expansion of proletarian suburbs and the rural exodus which provided much of the labor force needed in the new industries. At the same time, a part of the bourgeoisie fled the city, worrying about unsanitary conditions and unwilling to live together with people whose living standards were poor or even miserable. A new form of segregation emerges. The entire nineteenth century becomes obsessed with questions of hygiene: the prevention of epidemics, worry about water purity and waste disposal, air quality, sun exposure, children's cleanliness, and food safety in markets. Every political and

---

20. Sigfried Giedion, *Space, Time, and Architecture* (Cambridge, MA: Harvard University Press, 1982), p. 165.
21. Friedrich Engels, *The Condition of the Working Class in England*, orig. published in 1845 in Leipzig.

administrative manager had to make those requirements his priorities and so become more or less an urban planner. This cannot be reduced to the politics of taming and controlling bodies: everyone was motivated by new hope of reducing risks and improving sanitary conditions and comfort, which the new medical knowledge and new techniques promised and made desirable. Today we are still beneficiaries of these expectations. Another world is on the verge of being, and this world has an urban form. The multiplication during the whole century of utopian projects, some of which were realized, shows this very clearly. All of them, from Charles Fourier to Robert Owen, from William Morris to Victor Considérant, from Étienne Cabet to Tony Garnier, from Frank Lloyd Wright to Le Corbusier, propose to put an end to overpopulated, unsanitary, noisy, and dangerous cities.

Let us understand this reversal. Ancient, medieval, and classical cities were perceived as images of heaven. Today, images have emerged proposing to be the negation of real cities. We are facing a paradox: the city that was born as a monument rivaling the world of the gods and claiming to be the realization on earth of a potentially unified, organic, and balanced world now has become the instrument of its own displacement. The city seems to have forever lost its spiritual unity and symbolic form. The crisis started with the advent of a civilization that the city itself created and made triumphant. Hence the strange situation: the global success of the *urban form* seems to coincide with the *defeat of the city*, or at least with what for millennia was embodied and desired in the form of the city.

Let us conclude our reflection on the city as a machine by noting that it is important never to separate the city as monument, that is, as an architectural creation and architectonic image of the world, from its reality as an economic and technical phenomenon and a site of labor activities. As such, it is the space of the most radical social transformations of our history. It would be better to say that the city has been history's instrument par excellence in as far as there has been *too much history*.

The displacement of the city by the city itself makes it possible to understand the considerable architectural crisis between the beginning of the nineteenth century and the period of Modernism. The problem seems to have been as follows. The disintegration of the classic city was experienced as an invasion by outside forces, as if the triumph of the machine were intrinsically alien to the urban universe. The more pressing the danger, the stronger became the reaction to maintain the ideal of the traditional monument against the new "barbarism." Freestone apartment buildings multiplied with classical, neoclassical, or eclectic traditional motifs as if to exorcise and counter the triumphant machine industry. Architecture's refusal to use new materials (iron, cast-iron, glass, or concrete) for housing or public buildings (city halls, prefectures, and schools) meant that the city as monument wanted nothing to do with the city as machine.

However, as Giedion notes, new energies were allowed to express themselves and to prevail in sectors that did not concern the traditional idea of the city: railway stations, warehouses, indoor markets, covered markets, department stores, factories, slaughterhouses, exhibition halls, and suburban bridges. Here the engineer had to become an architect and in the nineteenth century most often became the inventor of new forms. But the increase of the engineer's importance was not sudden. It had been prepared mentally by the advances in the physical sciences and technological developments, as well as by the creation of engineering corps and schools such as Polytechnique in France in 1794. Just as significantly, the increase had also been anticipated in the last decades of the eighteenth century by a whole succession of so-called neoclassical architects[22] such as J. F. Blondel, J. G. Soufflot, E. L. Boullée, C. N. Ledoux, and J. L. N. Durand. They radically called into question the Baroque and Mannerist legacy and demanded that rationality in buildings be expressed according to two main registers: 1) architectural, by establishing a clear relation between

22. Anthony Vidler, *The Writing of the Walls: Architectural Theory in the Late Enlightenment* (Princeton: Princeton Architectural Press, 1987).

form and function and between the building elements and their technical necessity; and 2) social, by having the building meet the need that brought about its construction and inspire through its architecture the virtues expected from the public for whom it was built.

However, not much remained of that initial functionalism with its moralizing component, because this architecture still belonged to the monumental and epic city, magnificently glorified for the last time, even though underpinned by science and utility. Indeed the developing crisis goes far beyond the debate over the vocabulary of architectural forms and the question of their exemplary social nature. The challenge now concerns urban masses. New populations mobilized by conquering industry must now be housed. At first done without planning, construction was motivated solely by profit seeking, as for instance the "profitable buildings" mentioned by Balzac. One must admit that the proliferating suburbs of the industrial world were serious architectural failures, but nothing in the art of traditional building offered any creative solutions. The best architects so far had devoted their talents primarily to palaces, churches, and other public buildings. There seemed to be a total impasse between on one hand retaining the aesthetic norms of classical buildings that only rich neighborhoods demanded and could afford, and on the other hand the desire for innovative models whose construction forms would not project the cheapness of means conforming with revenues. Architects needed to innovate, use new materials, and also provide answers to large-scale social needs. Tony Garnier took on this triple challenge with his project *The Industrial City*, which he presented in 1904 and which was partially realized in Lyon. This exemplary effort contains both the promises and the ambiguities of Modernism. The promises consist in the stylistic renewal, that is, formal clarity and simplicity, new building trends (such as large windows and terraced roofs) and in the choice of materials (reinforced concrete, steel, and glass) and modest costs. The ambiguities appear in the urbanism that corresponds to a primarily functional rationality both in terms of movement (linear layout) and activities grouped in very distinct units (housing, leisure, stores, meeting or entertainment spaces, and manufacturing). This is not the kind of urban planning proposed by E. Howard's *Garden City*, nor F. L. Wright's *Broadacre City*, or A. Soria y Mata's *Linear City*, all well-known projects partially implemented at the time, but it prefigures Le Corbusier's *Radiant City*. While the latter invented marvelous forms, his plans for urbanism concentrated the pretensions and illusions of scientism and seemed to completely ignore the social and bodily aspects of the urban experience. This purely theoretical vision marred many avant-garde projects, including those of the Bauhaus, in spite of some exemplary architectural achievements. It privileged the representation of a universal individual, indifferent to local ways of living and traditions, and became known in the vulgate as the International Style,[23] whose latest avatar might well be the "generic city," promoted by Rem Koolhaas.[24]

However unacceptable Modernist views on urbanism may be, they nevertheless formulated the need for a formal renewal that would boldly and energetically assume the consequences and possibilities of the industrial age. Another world was emerging where urban space became the object of new experimentation in the treatment of volumes, and the creative use of materials such as concrete, steel, and glass, both in building structures and interior spaces. Furniture too, whose new norms have not been surpassed to this day, successfully achieved functional purity. But then the city entered an age of an even more critical transformation linked to the penetration of railways into city centers, automobile transportation, to gas lighting, and the telephone. Those technical networks transformed the circulation rhythms, communication, and all other forms of exchange. Another economic system and another urbanism were already emerging.

---

23. On the considerable work done on this issue, barely alluded to here, see Françoise Choay's text, "L'urbanisme en question," introduction to *L'urbanisme, utopies et réalités* (Paris: Seuil, 1975).

24. R. Koolhaas, *Delirious New York: A Retroactive Manifesto of Manhattan* (London: Academy Editions, 1978).

3

NETWORK: EXCHANGES, CIRCULATION, AND RELATIONS

We have established that the city as built environment has systematically tried to assume the form of a monumental body, although nothing prevents us from imagining it could have been otherwise. We have also understood that monumentality implied a large-scale technical system making the city into a mega-machine. However, we still need to understand that from the beginning, the city could only emerge as a *network*; better still, the city only really starts as a network and by developing as such, is transformed by it. This is probably the major aspect of the contemporary city, so much so that the question arises whether its monumentality and even traditional urban form might disappear. Such displacement would be very different from the one caused by the industrial revolution. We are no longer dealing with the city's anarchic expansion into zones of banishment, those poor and under-equipped suburbs where populations exploited by capitalist production are concentrated. This expansion is quite different and tends to produce archipelagos of urban sites. These may be comfortable but have no other goal than to provide essential services in contiguous areas of housing and offices. Have we gone from dislocation to delocation? Is this really the only thing that our most recent modernity has to show us? Aren't we also experiencing the invention of another relation to space after a four-century-old conception that unquestionably produced great works but could also be rigid and conventional, too solemn or too utilitarian? There are numerous signs of intense creativity and profound renewal. We are still undergoing this metamorphosis and the future city remains enigmatic. Furthermore, this upheaval acutely poses the question of public space whose traditional forms seem to disappear as monumentality associated with public space is challenged everywhere. Communication networks multiply new possibilities to exchange views and debates between citizens. This change is creating a new type of public sphere or even produces one in traditions that never considered it as an essential dimension of social and political life. The last part of this essay will deal with these questions. For now, it will be important to understand the implications of the city as a network.

THE NETWORK QUESTION

First of all, what is a network in general? Formally, we call a totality of interconnected points or terms a network. Those interconnections can be formalized in graphs. More precisely, a network is a system of relations where each term, through its links with neighboring terms, is virtually linked to all the others according to a saturation prerequisite and hence a system prerequisite. The local reaches the global through successive linkages. This description however is too general. As soon as we ask how a network functions, different types appear referring to specific uses. Their specificity can be better identified by looking at the etymology of the term since it condenses long-tested practical knowledge. *Literally* a network is a net (in Latin: *rete* or *reticulum*; in low Latin: *retiolum;* old French: *resel*). A net is defined as fabric with spaced intersections—links—a mesh that forms a grid of threads intersecting along two axes (*woof* and *warp*). It holds together because of the connections between each space, but especially because of the thread that forms the borders. A net serves as a light, airy, and supple instrument of capture. The various qualities of the *reticulum* have generated a large number of metaphoric extensions in several fields: 1) the circulation of persons and goods, in networks of roads, canals, railroads, and in sea and air traffic; 2) the transportation of energy or liquids in networks of gas, electricity, or water supplies; 3) networks of information transmission such as postal services, semaphores, telegraphs, the telephone, television, email, and the Internet; 4) systems of privileged relations between people and institutionalized groups in political, professional, religious, commercial, banking, and sports networks; and finally, 5) all forms of association and solidarity organized for legitimate or illegitimate purposes.

So many diverse uses of the term might seem excessive and leave little room for rigorous conceptualization, but there is a good reason to consider them all. It shows that etymology quickly ceases to be of any use. A network is not simply a grid or web and its concept must constructed by taking into account its genealogy. One initial distinction in the above-mentioned examples is obvious, between on one hand connective systems (streets, roads, railroads, and cables) and on the other hand circulation flows (pedestrians, cars, trains, and electricity). The first are called *topological* networks, the second, *rheological*, from the Greek *rheîn*, to flow. It seems difficult however to apply this distinction to social relations and therefore we must broaden the concept. It will be useful to briefly recall the origin of the term, linked to the history of mathematics and chemistry, electricity and geography. However, prior to this recent history, Leibniz needs to be credited for ideas centered on the idea of network, although he did not use the term. It is present everywhere in his mathematics, physics, combinatorial theory, monadology, and theodicy.[1] Hence his surprising modernity.

1. See Michel Serres' classic *The System of Leibniz* (Manchester, UK: Clinamen, 2001); see also R. J. Wilson, *Graph Theory 1736–1936* (Oxford: Clarendon Press, 1980).

Today's meaning of the term comes into use only at the beginning of the twentieth century. According to Briggs, Lloyd, and Wilson, in mathematics the concept of *net* emerges in the notion of *graph*, that is in a group of terms or vertices whose connections or boundaries are provided locally with a relational order. However, the concept of graph itself remained implicit until the end of the nineteenth century and first appeared in chemistry, in connection with problems of spatial representation of the atomic structure of molecules. Before that time, people used terms such as tree, grid pattern, web, ramifications, or more abstractly, combinations or analogous terms, depending on the field. In the last two decades of the nineteenth century, the term graph becomes explicit in the finite group theory of the English mathematician A. Cayley and later his compatriot, W. F. Clifford. At the same time in France, G. Brunel uses it to represent isomers. It becomes generalized among mathematicians at the beginning of the twentieth century and then merges with the term "network," as is shown in A. Sainte-Lagüe's work, *Les Réseaux (ou graphes)*.[2]

The concept of network becomes most important however in the domain of electricity, for obvious reasons. The problems posed by graph theory concerned the representation of certain types of relations. With electrical networks we go from the typological to the rheological, from abstract configuration to real flows. What is at stake between generator and receptor is the storage, transformation, and distribution of energy. If the system is on an industrial scale, it can entail considerable social and technical costs should it jam or collapse. Thus research in this domain became important, like the work done by G. R. Kirchhoff as early as 1845 on circuits, bifurcation, and the stability and linearity of electrical flow. Other research formulated problems of accountancy, maxima, conservation, and inversion. Even though less complex, problems of the distribution of gas and electricity (storage, distribution, layout and pressure) were no less technical.

2. A. Sainte-Lagüe, *Les Réseaux (ou graphes)* (Paris: Gauthier-Villars, 1926).

Geographers dealt very differently with the network question. Far from being peripheral to our problem, their approach is crucial since it concerns the organization and management of territories and the presence and activities of human settlements. Geographers ask very interesting questions such as: what is the relation between the centrality of a town and its multifunctionality, economic, administrative, educational, medical, or other? What is the average distance between a given town and a larger one in terms of maintaining an advantageous relationship? Is there an optimal distance? How do means of transportation and communication change the thresholds of this relationship? Is there a calculable link between a town's size and rank? Starting with such questions geographers have produced maps differentiated by subject, which look like mosaics of spiderwebs. Increasing concentrations of population are represented as dots of increasing size: villages, towns, small and larger cities, and regional metropolises. Multiple lines between points indicate the type of relations. In short, these are networks.[3]

The striking thing about their arrangement is a kind of *topological regularity*, which raises a host of questions, the most important of which is whether a specific network effect could account for the phenomenon. In 1863, an engineer of the *Ponts et Chaussées*, Léon Lalane, was the first to develop a theory based on the study of a large number of agglomerations distributed according to three levels or orders: small, medium, and large. He concluded: "The average distance separating neighboring agglomerations of the same order is the exact multiple of the medium distance separating those of the lower orders."[4] This surprising conclusion is followed by another: when these distances are represented by triangles, their design shows "a pronounced tendency to come close to an equilateral triangle." This presupposes the existence of a

3. Philippe and Geneviève Pinchemel, *La Face de la terre* (Paris: Armand Colin, 1988), part II.
4. Ibid., p. 84.

powerful and unconscious topological schema capable of regulating the distribution of urban settlements. It is a bold approach.

In 1930 Walter Christaller conducts a similar investigation for southern Germany; he discovers that the dominant form of the representation of relations between cities, according to a centrality gradient, tends to be a regular hexagon. This figure is known to generate a system of juxtaposition without gaps, as in the case of the square or equilateral triangle, while offering an optimal number of connections. This resembles the cells of a hive. Christaller proposes different models according to the types of activities: commerce, transportation, and administration.

In 1949, the American sociologist G. K. Zipf demonstrated the existence of a constant relation between size and rank, which model continues to be refined. These pioneering approaches were taken up again in the sixties and today they make it possible to draw complex maps of relations between metropolises, between regions and metropolises, and also between urban poles on a global scale. Other analytic instruments were developed, based on statistical data, to establish either the curve representing the increasing complexity—connectivity index—or the number of nodes in a given space—density index—or the relationship between the number of nodes and the number of connections. The mathematical formulation of these indices permits simulations or comparisons between different types of networks. While the results should always be considered part of a comprehensive approach, they are revealing. Another revealing approach replaces the representation of distances by travel time, sometimes associated with costs. Such maps are surprising from a territorial perspective but are consistent as networks; significantly for instance, the erasure of territory between two cities as a result of the speed of the means of transportation is called the "tunnel effect." Topological and statistical approaches have the advantage of revealing tendencies and constants that seemed accidental at the level of perception. They offer a remarkable change of scale and a more distanced and quasi-geological view of populations and their habitats.

Urban planners might be surprised by one aspect of the geographical approach to urban networks: their studies deal primarily with the relations between cities, not with the cities themselves. But this is precisely our question: *how does a city as such constitute a network?* What does this add to its other dimensions? How does this throw new light on the issue? Our brief summary of the network problematic in the last two centuries may at least have raised the expectation that the new emerging urban space will primarily have a reticular logic, both at the level of social relations and systems of production and exchange. But how does this affect architectural design itself? In order to answer that question we will need to highlight a number of general *formal characteristics* of networks that can apply to other wholes, starting with urban space. Let us outline a typology of the main traits of what can be called a network.

1. The first trait could be defined as decentering or rather multicentrality. This appears to be the most obvious characteristic of any network and constitutes its epistemological originality: there are as many centers as there are nodes. With respect to the physics and metaphysics of the fixed point, a network invites one to think in terms of groups of relations. It allows for multiple points of entry because the reference is the totality of connections itself rather than a single position, even if subgroups inside the network can seem important or dominant. In terms of urban circulation, it can be understood as follows. In an orthogonal layout, as in ancient Greek, Roman, Chinese, and most modern cities, starting from any point and giving access to a minimum of two lines intersecting at a right angle, one can join another point on a different line without passing through any one specific interconnection. This is the contrary of a star-like arrangement without lateral links, so-called "glove fingers." This means freedom of movement within the network or, at another level, the possibility or promise of flexible social relations, as opposed to a hierarchical model, and hence the valorization of *vicinality*. Thus the reticular arrangement of streets runs counter to monumentality.

2. The second trait can be described as the *coherence* or *interdependency* of elements resulting in their coordination. Coherence, because by definition a network consists in mutually linked points that can be represented in one graph or a group of graphs. Terms or points only exist in relation, like stitches maintained by reciprocal links. Concretely this poses the problem of *interconnection,* as in electrical and electronic connections. Coherence is then the effect of local connections even if this does not exclude a central control. Socially, in urban space this translates into the importance of solidarity between neighbors or within a neighborhood. At a deeper level this gives rise to the formation of groups linked by a specific project, an ideology, or a belief. In associations for instance, solidarity appears as *coordination*, the effectiveness and rapidity of intervention in emergencies, protests, celebrations, and fund-raising.

3. Let us call the third trait *openness* or the capacity for network extension. This means the possibility of constantly developing new connections, either by tightening the links or enlarging the perimeter. Because the system is not centered, there is never any imbalance with respect to one single reference point. Architecturally this means that one proportional standard can be repeated without upsetting the scale. The advantage is that the size of the whole can change without breaking its formal equilibrium.

This appears to be one of the essential requirements of contemporary urbanism, as it was for Walter Gropius for instance. Decentering allows for the extension of through-linkable units or reproducible modules. The new urban spirit makes it possible for mobile populations to accept new settlements only if they appear to fit in the network and not as fringes, or outside a central space seen as privileged or exclusive. Obviously looming here is the question of the success or failure of new cities.

4. Fourth, let us call *specification* or *particularization* the network's capacity to constitute niches of relations inside a whole or to form autonomous and *legitimate* subsets; it seems that where nothing is central, nothing is peripheral either. This is clear from the previous characteristics but there is an additional dimension in

the case of the city. Not only is a network capable of creating specific spaces that are locally organic and globally connected, but it can include the most varied modes of living and traditions. This has always been true of cities. We still see this in any city where local groups are formed on the basis of professional, ethnic, or religious connections, or arise out of specific projects such as clubs and associations. These local groupings are added to the whole without being considered marginal. The network structure tends to multiply centers that appear as urban relays valorizing the variety and singularity of neighborhoods. This is a *distributional* model that reflects the possibilities of openness, decentering, and interconnection mentioned above.

5. Fifth, let us define one of the major qualities of any network as *accessibility*, which means that it is possible to access any point from any other, and also that no one has the right to block such access. Whoever claims this right will face strong challenges, because it is in the nature of networks to be accessible. The Internet is a dazzling example. In urban spaces such accessibility takes on additional meaning because the city itself becomes the space where the most varied interconnections cross or are superimposed without centralization. The city is a network of networks. Hence the paradox: it appears as a center of decentering. And so the city offers access to local, national, and international transportation; it favors access to communication networks, knowledge and cultural sites, health centers, and administrative services. Finally, the urban site is marked by a general *availability* that favors random encounters in the open spaces of communal life (streets, restaurants, cafés, shops, and entertainment places), while preserving the freedom to abstain, in other words, to choose whether or not to activate relations.

6. Sixth, finally, the most distinctive characteristic that necessarily defines the reality of a network is *mobility*. It is not just a possibility available to circulating flows or connected agents. More essentially, it is what networks imply and offer, as opposed to a centered system, by the very fact of its decentering and the multiplicity of their nodes. If a network is necessarily multicentered,

then no stable position is guaranteed within it, precisely because its function and power are due to its ability to liberate movement, to enable flexibility, to adapt and innovate, to extend, and finally, to have the malleability or capacity to renew its program. Mobility is at the heart of many urban questions. Since the city is from the start linked to exchange and by definition a system of capturing and launching various forms of flow, the fluidity of movement is essential. Even with ramparts, the city is virtually open. Only war can close it, but even then for a limited time. A city where mobility is stifled is in a sense besieged. A city reaches its optimal size through its belonging to a vast network of connections to other places and other cities. It is a port at the crossing of roads, railroads, and air traffic, and sometimes rivers or canals. Those are the conditions for mobility in urban settings that can contribute to social mobility.

NETWORKS AND URBAN SPACE

We could probably identify other pertinent characteristics of networks[5] but these are sufficient to further develop our approach to the city. But first, there are two preliminary questions. The first is to ask whether applying a network concept based on recent mathematical, technical, and sociological formulations, to ancient and even very ancient cities, might not project a contemporary problem onto the past. The second is to consider whether even in the case of the most ancient cities the concept can refer to a decisive transformation with regard to the two dimensions mentioned earlier, the city as monument and machine. It is easy to see that the objection to the first, semantic, question, does not hold. The con-

5. There is not yet much work done by urban planners and sociologists on the network model and its historical metamorphoses, but it is promising (cf. G. Dupuy, M. Castells). What needs to be investigated is whether it is possible to rigorously represent observable data with formal models like those in mathematics and the theory of graphs. My remarks here are thus completely exploratory. See also S. Marvin and S. Graham, *Splintering Urbanism: Networked Infrastructures, Technological Mobilities and Urban Condition* (London: Routledge, 2001).

cept of "civilization" which we use for all epochs, appeared in its present meaning only in the second half of the eighteenth century. More important, however, to answer the second, epistemological, question, we need only to remember that we can read Greek physics with our concepts without committing an anachronism. Today's terms express yesterday's problems. It cannot be otherwise.

It should be emphasized that the so-called checkerboard building model, the orthogonal grid, appears in cultures very distant from each other in time and space: Sumer, Etruria, and Rome, China, Greece, and Tenochtitlan, despite great differences in architectural organization, style, and function. Nevertheless, the structure of intersecting lines or succession of regular blocks is conducive to original housing forms, ways of living, and social practices. Rome remains a good example. The model of the orthogonal grid organizes not only the religious, political, and technical dimensions of the city, but also farmland, according to the same science of territorial measurement called *planimetry*; it makes possible a cadastral map of inhabited sites. Paved roads establish contacts among cities and among provinces. The empire is a network, and even a network of networks.

We can now ask the question again: how relevant is the network model for understanding the city in general and in particular why was this dimension somehow released in contemporary urban space? Every city faces the same task: to concentrate in the same built environment a population that must get organized, create institutions, reside, communicate, circulate, feed itself, and find water and fuel; produce goods and transport them; evacuate waste, resist illness, teach, educate, live in peace; defend itself militarily, and develop. The grid, that is, the mini-max type, is the best answer to relations between flows and positions, random movements and fixed spaces, population density and circulation. This leads to another question: does the network have only virtuous and no perverse effects? That is highly questionable.

Let us look at the case of Mesopotamia and the formation of cities in river valleys. As we have seen, for more than a millennium agriculture developed in the hills of the Fertile Crescent before the emergence of the city proper. The first urban manifestations are very limited. However, everything speeds up when farmers settle close to the rivers. They benefit from the rich soil and abundant water thanks to the introduction of canals, hence they depend on specifically technical activities. Something has changed: it becomes necessary to import what the valley does not provide, wood, stone, and metals, often from far away. The appearance of the city cannot be separated from exchanges with the outside world, which from the outset implies a network of relationships with groups of suppliers, and a circulation of people among these various groups. Those are political as much as commercial relations. This is unlike what happens in villages where networks, as in all traditional societies, are defined by kinship systems or personal allegiance and where exchanges remain local. The city presupposes and develops open network relations with partners that are not defined by ethnicity, although this is often the case, but by their function and the services they render. One could say that the city arises through this circulation and stimulates it. This concerns exchanges of goods but also political and military alliances. The city lives through a network of exchanges and ensures its security through a network of alliances. This may lead to confederations such as the league formed around the powerful city of Athens in ancient Greece after the victory over the Persians. Often this is how empires arise, as was the case in Mesopotamia, China, and Rome.

The network model, resulting in new relations between interior and exterior that mark the emergence of cities, must be understood at the deeper level of the city's operation and the transformation of its identity. This can happen in at least three different areas: 1) in the urban space itself, 2) in the administrative system, and finally, 3) in the circulation of information and knowledge. Let us look at those three points.

Wherever it can be observed historically, monumental urban space is the most public and forceful manifestation of power. We have seen that in Mesopotamia the difference in the size of the buildings and their hierarchical differentiation was the defining feature of urban sites as opposed to large peasant villages. In the Chinese city model, the entire city and its walls become the holy site par excellence, the noble space as opposed to the countryside. The city itself establishes an immediate hierarchy between temples, the prince's palace, the nobility's residences, and the houses of the people. The same can be said of Rome and most cities in Western antiquity, even if hierarchical separations are less coded. The city as monument is the immediate spatial embodiment of power, of orders and status.

Interestingly however, one of the city's elements transgresses hierarchical separations, and that is the *street.* Familiar as it is, the street is a surprising urban element: it is not a monument but a crevice, a space for transit, and also a space for encounters and more officially, a market site. In fact, its function differs greatly from one civilization to another and one epoch to another. In Mesopotamian as in ancient Chinese cities, the street is just a kind of circulation corridor between residences surrounded by walls. In both cases however commercial life sprung up spontaneously under the walls. In Greek or Roman cities it was customary for artisans to have shops opening up onto the street. The streets where citizens of all social ranks circulate, meet, and trade also have to be used by social and political elites to move around or go home. Not only do streets mix the most diverse populations, but they also give access to all the monuments and palaces of the city, or close to them. From any street you can reach any location in the city. Streets cannot be closed off because the backs of the houses border on another street. You have to close off an entire block or neighborhood. We have seen this in ghettos or cordoned-off neighborhoods where the network logic then functions again within the enclosure itself. The grid or other street pattern constitutes a network that subverts the monumental hierarchical order spatially and socially. The spatial existence of the street is a decentered network with undetermined uses and connections, and contains the seeds of one of the city's characteristics, the possibility of egalitarian relations, exemplified by the "municipal freedom" of the Middle Ages in Europe. Beyond this urban planning approach to the street, we will later see its importance, associated with the square, as a privileged example of *common space.*

We also find the network effect on monumentality and its social stratification in another domain, that of *administrative organization.* It serves a central power even if its main objective is to ensure the functioning of the whole society. The operational mode of any bureaucracy is hierarchical. It is a one-way system of interlocking powers in which responsibility always covers the lower ranks and where orders go from the center to the periphery. This also applies to knowledge of all the plans and access to files. In such a situation, another system develops through the interaction of urban relations that become reversible and multipolar, as in commercial exchanges, craft, responsibility for neighborhood problems, feasts, relations between generations, and the reception of visitors. From a political perspective, this network of relations invents what later will be called *civil society*, which sociologically should be broadened to include the concept of *common space.* This is a typical urban phenomenon because it transcends or minimizes kinship relations, ethnic or territorial membership. It valorizes professional membership (artisans, merchants, and clerks), which has played a decisive role in struggles for community freedoms, a prelude to the vast democratic movements of the nineteenth century.

The network effect is also played out in the acquisition of competencies and related powers. In traditional societies, competency is acquired progressively as one grows up, in hunting, gathering, planting, animal husbandry, or the mastery of artisanal techniques and warrior arts. Transmission takes place from adults to the young. It accompanies the growth of new generations. Very early on, a new type of competency appears in urban systems and their management that trumps all others, that is, administrative competency. In most cases (Sumer, China, Egypt) competency is linked to the mastery of writing. This confers considerable social advantages over all other professions. The urban site that created the conditions for the emergence of writing also tends to favor the movement of knowledge and know-how, and to make writing accessible to noncivil servants and to create independent training centers. Thanks in particular to the much easier-to-learn phonetic alphabet, this is the case in sixth- and fifth-century Greece[6] in the numerous philosophical schools where the Sophists represent an important, and specifically urban, movement. It can also be seen in European cities from the twelfth century in the effervescent creation of centers of learning leading to the birth of universities.[7] In the latter case also, the network logic born in urban spaces prevailed imperceptibly but effectively over the rigid positions of certain sectors of the church *magisterium*.

These are only some of the network effects that could be analyzed at the spatial, administrative, social, or cultural level. They are present at every material physical level of the urban system:[8] roads, water supply, wastewater canalization, and since the end of the nineteenth century, delivery of energy, electricity, and gas, and finally, distance communication networks, telephone and optic cables, and cell phone networks. We are beginning to realize how much all those networks created by the city have changed its ancient face. The first networks of circulation and exchange of goods were the conditions for the emergence of urban reality while the networks that created communal life and scholarly culture established the city's political power. Today the networks that are spreading worldwide seem to express the triumph of this model born with the cities.

However, just as the industrial revolution destroyed the organic unity of the traditional city, giving rise to the suburban phenomenon that remains an unsolved problem in modern societies, similarly the exponential growth of communication networks ends up calling into question the very idea of the city. For a second time, although in a different manner, a process created and accelerated

---

6. E. Havelock, *Origins of Western Literacy* (Toronto: Ontario Institute for Studies in Education, 1974).

7. J. Le Goff, *Intellectuals in the Middle Ages* (Cambridge, MA: Blackwell, 1993).

8. See G. Dupuy, *L'urbanisme des réseaux* (Paris: A. Colin, 1991), and Manuel Castells' synthesis, *The Rise of the Network Society* (Malden, MA: Blackwell, 1996).

by urban space results not only in making the traditional city obsolete but even *the city as such*. For the power of the networks has become so important, their ubiquity so total, that, in principle at least, there are no privileged sites.[9] According to some visionaries, communication networks are more and more indifferent to geography, including urban geography. Centers of production, management, and decision-making can in principle be linked to a city or not. They can move from a well-known place to an obscure one without their operations being affected. Only their agents' social life would be affected but it is more and more obvious that they are indifferent to typically urban sociality. Are we dealing here with the "end of the city," so often foretold? Is this really what is happening? It may not be the case at all for reasons that are due to the logic of networks. We will return to this later.

If indeed the very idea of public space emerged with the city, at least in the West, are we witnessing also the end of that space? Or are we dealing with the beginning of an inevitable transformation? But if so, which one? Similarly, how can we understand that the ideal of shared living in limited spaces, which has always been the originality of typical urban sociality, seems to disappear today or is only manifest ephemerally and sometimes exultingly in sports arenas, concert halls, political meetings, religious ceremonies, or mass assemblies? What remains of neighborhood life? In short, do we still understand what "*a common world*" means? These questions remain to be discussed.

NOTE: MONUMENT, MACHINE, NETWORK, AND THREE ENGINES

Before starting this debate, let us return to the three terms that have guided our first approach. We understood these to be three articulated modes in which the city emerges, develops, and

---

9. See M. Castells, *The Rise of the Network Society*, chapter 6, "The Space of Flows."

changes. Those three modes coexist and cannot be dissociated. We are dealing with synchronic modes and not stages. Nevertheless, it is as if time unfolded those dimensions one after the other as a result of changes that affected the whole society. What should intrigue us is the remarkable homology between these modes and the three types of engines recorded in our history: the vectorial, the transformational, and the informational models.[10] This correspondence throws an original light on the configuration.

The vectorial engine is the mechanical motor we have known since the simplest machine: fulcrum, lever, and exterior force. This force can be muscular, animal or human, or extracted from the environment, wind, water, gravity, or come out of material properties such as elasticity. The fulcrum and lever couple may be more complex (wheels, screws, pulleys, counterweights, or brakes); nevertheless the machine amounts to a system that captures power to create movement which, applied to an object, becomes work: mill, spinning wheel, plow, winnower, churn, crane, cart, and clock. The system is homeostatic: it intercepts power and applies it. It is a stable order where time is reversible. Even when machines become complicated and the natural is invaded by the artificial, we do not leave the system of displaced power. The mega-machine city, no matter how powerful, remains in the world order and confirms it. The mega-machine is equal to the monument and hides behind it. However, even when animal or human power requires a costly renewal of energy, when more fire is needed for the forge, this remains marginal in terms of a global equilibrium. Like the universe from Ptolemy to Galileo or from Newton to Laplace, the city may modify its organization but it nevertheless remains a stable configuration. It remains so until the emergence of the second engine. This means that for several millennia after its first emergence the city did not experience any global variations until just before the industrial revolution.

---

10. I am indebted here to Michel Serres' formulation in his stimulating text: "Moteurs" in *Hermès IV, La distribution* (Paris: Minuit, 1977).

Everything changes with thermodynamics, with the transformational engine, concretely with the steam engine.[11] In other words, the vectorial engine that transforms movement, such as the paddleboat or the weaving loom, may remain the same. However, the power moving it is no longer captured but produced by another engine that is operated by the heat accumulated in the boiler; the latter generates pressure through the difference between two sources. This force comes from consumed *energy*, produced by burning vegetal and mineral resources. No energy is released without this fire and destruction. As Serres has noted, we no longer deal with a transfer of power but with a *change in state.* Both the process and the loss are irreversible. We then discover that solar fire had a beginning and will have an end. The universe no longer lives in the perpetual movement of vectorial mechanics but has fallen into time ruled by entropy, according to Carnot's second principle. This changes the very idea of history. We must draw from the reservoir, but is it inexhaustible? We realize that all energy has a cost, including human labor which is now part of the balance sheet with other forms of energy. Marx implicitly understood this. Like the universe, the entire social order is now part of the furnace model. The same is true for the city that came out of the industrial revolution.

However, does it also mean that the city has become a thermodynamic machine? This makes no sense unless we say that it has always been the case since the classic mega-machine with its lever and pulley functioned with human or animal energy; however, this connection was not taken into account. The exploitation of the biotope was considered part of the natural order; hence the indifference of philosophers in antiquity to slavery. Thermodynamics has taught us that human labor itself presupposes a form of energy whose renewal has a cost. The corollary is that the machine also works and devours energy like a living being. This is a very differ-

11. The steam machine is not new; it has an ancestor in the eolipile of Heron of Alexandria (first century AD) and was reinvented in the seventeenth century. What is new in the eighteenth century with T. Newcomen, J. Watt, and their successors are the technical improvements and industrial applications.

ent perspective which corresponds to a considerable change in the scale of production. With the arrival of the steam engine, followed by the internal combustion engine and electrical turbines, industries no longer depend on natural forces such as water or wind, but they settle as closely as possible to the labor supply in the cities, and then attract and concentrate new populations. The city becomes the site of the exponential growth of industrial production, which is proportional to the no-less-exponential growth of energy consumption. This can be measured by the demographic explosion: between the beginning and the end of the nineteenth century, for instance, London's population goes from 800,000 to four million inhabitants. The dismantling of the ancient monumental model is visible as much in the disorderly proliferation of suburbs as in the architectural eclecticism betraying the loss of classical points of reference.

At the time the awareness of the existence of entropy inherent in the thermodynamic machine does not seem to have depressed the people. The prosperity it created and the promise of unlimited techno-scientific growth were enough to eliminate any negative outlook. The reservoir seemed inexhaustible and the future of science necessarily glowing. However, something else was happening: the powerful rise of the third generation engine, the informational engine. Clearly, energy itself is coded; software runs the hardware, which capacity must be read as something that ascends the entropic slope, as for instance multiple connections between synapses compensate for the loss of neurons in the brain. In short, the thermodynamic age almost immediately changed into the information age, which would become the age of the global expansion of networks. This is where we are now.

We must return to the questions we asked before: what is happening to our urban space? How does the crisis it has been undergoing affect our collective existence?

Second Approach

Rethinking Public Space, Discovering Common Space

4

THE DOUBLE MEANING OF PUBLIC SPACE

The idea that the city is the primary site of public space seems foremost in Western thought. Apart from the question of the origins of democracy in Greece and Rome, there is an essential reason for this conviction: the continual use of a political vocabulary in many European languages which merges with the vocabulary concerning the city. In Latin the term *civis* citizen comes straight from *civitas*,[1] the city. In Rome, the citizen is the city dweller, even as the city included the countryside and each citizen remained linked to a *gens*, a lineage that could extend to families far removed from the urban center.[2] In Greece's Attica, after Cleisthenes' reform in the fifth century BC, each of the ten tribes forming the polis included a truly urban population, another situated on the coast, and a third called inland people. In Athens particularly, the architectural design and arrangement of the buildings made it clear that the city space was intended to trans-

---

1. This is a complex relation, as É. Benveniste explains: "In Latin, *civis* is the primary and *civitas* the secondary term." *Problems in General Linguistics*, translated by Mary Elizabeth Meek (Coral Gables, Florida: University of Miami, 1971).

2. Such membership remains a permanent feature in all ancient cities in Europe, the Middle East, and China. However, according to Weber, the European city, at least in the West, clearly favors professional relations over lineage membership (cf. M. Weber, *The City*, chapter 2, "The Western City").

late visually the fact that power is defined (if not always exercised) according to a democratic model. Although it is too often idealized, Greek public space remains a model for us. It is the space where the members of the city, following agreed-upon rules, debate and decide in an open forum. Public means that which is exposed to the community's view, judgment, and approval. We now understand the concept of "public" to mean the indissoluble bond between visibility, transparency, and citizen judgment. This explains why the notion of public space evolves from its urban origin, the built environment, to the idea of an institutional system of open and contradictory debate. The objective is to achieve reasoned agreement between members of the city on the definition of institutions, the formulation of laws and their implementation. From that perspective, public simultaneously means open to all, known by all, and recognized by all. Public space is the civic space of the Common Good as opposed to the private space of particular interests. In Western thought this conception has become the basis for the normative evaluation of what can or must be understood as public space.[3] The global expansion of this model of democracy tends to confine the debate in these terms. However, this obvious assumption needs to be called into question; the concept of the public sphere must be broadened and, more important, differentiated. Such a critical analysis will perhaps give us a different perspective on the crisis of public space faced by contemporary cities. We need to ask, what *space* are we talking about? Clearly we are dealing with a metaphoric phrase in which "space" designates a system of relations governed by certain norms. By acknowledging this, we can avoid empty wordplay.

Can we understand "public" in a different sense from the meaning that has prevailed in the political tradition inherited from European antiquity and revived by Enlightenment rationality?[4]

---

3. H. Arendt, *The Human Condition* (Chicago: University of Chicago Press, 1958). See also M. Hénaff & T. Strong, *Public Space and Democracy* (Minneapolis: University of Minnesota Press, 2001).
4. J. Habermas, *The Structural Transformation of the Public Sphere* (Cambridge, MA: MIT Press, 1962).

Possibly. We saw in our discussion of the emergence of the first cities in Mesopotamia that one of the essential criteria applied by archaeologists to call a settlement a city was the existence of collective constructions such as ramparts, temples, and palaces. To qualify them as collective does not mean that they were accessible to all, but indicates that they were not private dwellings. Even if the palace was in fact the residence of the prince or the lord, it was different from a dwelling. It was the site where authority resides and is exercised; its distinctive, monumental, and richly decorated architecture was designed to signify the grandeur and transcendental or dominant character of power. The same can be said of the temple dedicated to a deity. The city walls not only guarantee its security but also express its power and outline a dedicated space. These constructions can be seen by all even if in some cases like palaces they are separated by walls that signal the distant presence of sovereign authority. The visibility and monumentality of religious or princely buildings cannot be compared to the transparency or sharing of common space according to the norms of Greek democracy. Nevertheless, these buildings are part of the city and can be seen by everybody. This monumental visibility should be called public. In any case, it is not exclusive to ancient non-Western civilizations with more or less despotic power structures. On the contrary, it also corresponded to the most common experience of urban civilization in Europe itself. Princely palaces were just as majestic and distant. Their grandeur was there to be seen. Let us add that republican palaces and governmental buildings in modern democracies are no less difficult to access and remain just as imposing as those of ancient monarchies.

These remarks suffice to make a distinction between two concepts of public space. The first concerns space offered to the gaze of the public: it defines sites of *common visibility*. These may be buildings where sovereign authority is located, religious buildings, museums, hospitals, universities, but also banks, company headquarters, or department stores. The important thing is for a building to participate in the life of the city as a monument by its size and the quality of its design, as space marked by its architecture to signal it is more than a dwelling. Monumental visibility cannot be separated from the urban landscape as a whole with its avenues, squares, streets, and parks and constitutes a fundamental signifier of public space. This does not depend on whether the society in question is democratic or not.[5]

This public aspect should not be linked to the specifically political concept of public space where the word space itself takes on a metaphoric meaning. "Public space" must be understood as an institutional system of democratic debates and decisions that take place in distinct sites such as the parliament, the presidential residence, and government departments. The fact that public debates and the exercise of power were linked to one type of spatial organization in the case of Rome and Greece does not mean that democratic debate is dependent on a single spatial organization or privileged urban form. Historical or ethnographic research shows great variety in assembly practices, for example in Gallic or Germanic tribes in Caesar's time, as well as in Japanese peasant communities of the Edo period or the agricultural villages of the Ochollo in Ethiopia in the twentieth century.[6]

We can therefore affirm the difference between two concepts of public space. They are not divergent but parallel. One concerns monumental visibility in urban space and had traditionally a privileged relation to the architectural expression of sovereignty and centralized administration. The other concept defines the sphere of public debates and may or may not be linked to spatial expressions that conform to the political concept. However, in our recent

---

5. The case of China is very instructive. According to F. Billeter we cannot speak of politics in China in the sense of our *politeia*. The main reason goes back to a millennium before our era in the Zhou dynasty, which established a system where power relations were translated in terms of a kinship hierarchy. This introduced rare stability in those relations. There were other consequences however: "Our notions of the State and the 'political,' the opposition between 'public' and 'private' have no place in China or do not have the same meaning" *Chine trois fois muette* (Paris: Allia, 2006), p. 105. Furthermore, power does not refer to a legal order but is defined by the strategic ability to influence and persuade in a framework of relations that have never ceased to be hierarchical (cf., ibid., p. 116 sq).

6. See Marc Abelès, *Le lieu du politique* (Paris: Société d'Ethnographie, 1984); see the very informative work, under the direction of M. Detienne, *Qui veut prendre la parole* (Paris: Seuil, 2003).

modernity characterized by the expansion of networks, both conceptions of public space are profoundly called into question.

5

FIRST PROBLEM: THE CRISIS OF THE MONUMENTAL MODEL

Today the first conception is questioned in three areas: technical, or the concentration of productive forces; managerial or administrative centralization; and finally, architectural, or monumentality itself. Let us rapidly review the changes.

Industry in the nineteenth century tried to take advantage of the urban machine as much as possible, from its capacity to concentrate the labor force, to manage large populations (to house, register, supervise, classify, police, provide supplies, educate, protect, maintain, and prevent.)[1] Industry chose to locate closer to sites with a concentrated labor force as soon as energy (coal, oil, and electricity) could be moved. The industrial revolution was first and foremost urban. However, we cannot identify the city with industry and Western capitalism with the urban phenomenon alone. The city is neither industry nor capital but it has been the technical and social condition for the emergence of all technological and industrial revolutions. We have now arrived at a paradoxical reversal where modern industry no longer needs the city, espe-

---

1. M. Foucault understood the novelty and scope of the exercise of control from the confinement of high-risk groups to the hygiene problem of the whole population. See: *History of Madness*, *Discipline and Punish*, and his course on bio-politics (1977–1979).

cially now that the city has become unlivable and the inconveniences drawbacks that have outstripped the advantages. Industry has started to leave the cities behind, often in a state of irreversible decay.

NETWORKS FRAGMENT MONUMENTS

Today, communication and transportation allow for a complete decoupling of the city and the factory. Industries settle wherever they like, taking into account for instance the price of land, access to transportation, storage, or legal and fiscal advantages. Their location in turn determines housing developments. It leads to urban archipelagos without a city, a random sequence cluster of agglomerations consisting of a mixture of industrial zones coupled with housing developments where commercial centers and public services such as police stations, fire stations, local administration, schools, and health centers often later emerge.

The city permanently loses the ancient function of mega-machine, in other words ceases to be a system for organization and production *as such*. The city is not alone in managing, distributing, and granting status; another authority asserts itself which is the network of industrial organizations in variable locales. The system seems to become completely dematerialized. What was contained in administrative files is now carried in electronic databases. The city is no longer the context that guarantees an available workforce, which can come from anywhere and is in fact available anywhere. The city is no longer the geographic and social site where it settles. However, we will see later on that for other reasons the city will preserve and even renew its ability to remain a management center.

We must admit that the ancient technical and social functions have been transferred for the most part to communication networks. Does this mean that the built environment and the population concentration it presupposes are no longer important? Many scholars claim that we are in the process of going from a purely spatial and volumetric arrangement to an ubiquitous and immaterial system, even though it rests on material underpinnings. The industrial revolution had obliged us to reinterpret the city's future by revealing the technological dimension of the machine implied in the architecture of the city as monument. Similarly, the electronic revolution, which generalizes and globalizes communication networks by radically modifying industrial space, forces us to reinterpret another dimension which had been present from the beginning of the urban phenomenon, that of circulation and exchange; in short, *networks*.

This brings us to a second point: the obsolescence of centralized administration. Like some urban planners, we wonder if capitals will still exist at the end of the twenty-first century. Of course, national and regional metropolises will not be eliminated. They will probably remain what they are today, privileged places where political, economic, and administrative decision makers reside. However, we are told that such geographic settlement could become more and more symbolic. Many administrative services are already scattered around the nation. This tendency is striking in a country like France which was highly centralized. In other cases, like the United States or Germany, the process had started a long time ago. From the beginning, the United States has known that the administrative capital need not be identified with the economic or cultural capital. This is true for Washington, but also for most of the state capitals (see Albany for the state of New York, Sacramento for the state of California, and Baton Rouge for Louisiana, etc.). The need for political and symbolic centrality is unraveling. As we know, in reality the decision making center moves with the representative. Heads of state today can travel without being accused of deserting their post. There will be no more flights to Varennes.[2] The symbolic value of the capital city still seems strong but can it resist the erasure of its former organizing and centralizing function? Or can a different dynamics give it another chance?

2. [Note from the translator] This refers to the attempted flight by Louis XVI to join the enemy troops. He was arrested by the revolutionaries in 1791.

In ancient civilizations power was associated with the concentration in one place of all the instruments of domination, influence, or organization such as weapons, money, messaging services, offices, knowledge, and production. Those who held supplies also owned the world. This "monumental" concept of power has become obsolete. "Today," says Michel Serres, "relations between medium and transportation are turned upside down as the latter becomes essential, as always. Storage sites no longer matter because our networks connect them and they can thus be as scattered as the nodes that exchange information."[3] Today supplies are flows. Flows stream through singular network sites, establish communication between them, and raise them to the global level by circulating through the whole system. However, those singular sites resist; without those stable points the flows themselves could well lose their operational power. This is more and more apparent.

The third aspect concerns the loss of monumentality itself, or rather, the end of monumental public space. All monuments from ancient civilizations publicly express political or religious power. According to Georges Bataille: "Large monuments rise up like barriers that oppose the logic of majesty and authority to all the shady elements; in the shape of cathedrals and palaces the Church and the State speak to the multitudes and silence them. It is indeed obvious that monuments inspire good social behavior and often even fear. The capture of the Bastille symbolizes this state of affairs: it is difficult to explain the crowd's action other than by the hostility of the people against monuments that are the true masters."[4]

We should guard against assuming hidden motivations, as Bataille tends to do, that would reduce the grandeur and power of impressive ancient monuments to the sole purpose of creating formidable displays of domination. In both the Greek and Roman tradition, the splendor and monumentality of temples, palaces,

---

3. M. Serres, *Atlas* (Paris: Bourin, 1993), p.152–153.
4. G. Bataille, "Architecture" in *Documents* 2 (May 1929), *O.C.* Paris, Gallimard, 1974, t.1, p. 171.

theatres, and other important buildings are attributed first of all to what they represent: the majesty of the public domain. This was so well understood and inscribed in stone that those majestic forms have survived till today.

However, this is precisely where we find the architectural crisis. On one hand, power no longer needs monumentality to assert its legitimacy, even if former palaces continue to house the nation's elected officials and the state's civil servants; on the other hand monumentality is on a different scale and has a different function. A skyscraper housing ministerial offices has nothing in common with a church, as we see paradoxically in the case of St. Patrick's Cathedral in New York, jammed between buildings that are about a hundred meters higher.[5] The skyscraper does not belong to the category of monument, not only because it is primarily utilitarian compared to a temple, a museum, or triumphal arch, but because it is no longer proportional to human space. It is no longer a creation that emerges from the city to express its glory; it seems to have come from somewhere else. It is outsized, which is the definition of colossal, and it is like a giant object posed on the ground. A neighborhood of skyscrapers is a section of the city projected vertically, where the elevators function like subways or buses. The territory is in a way tilted toward the sky. This is why an avenue in Manhattan is nothing like an avenue in Paris, London, or Rome; it looks more like a man-made canyon, a monumental hole dug vertically into urban volumes. Such an idea might occur to anyone flying over New York on a bright morning, making the city look like a dark sandstone massif sawed into parallel blocks by a modern titan.

While private buildings are becoming "monumental," many new public buildings, like city halls, law courts, schools, and even churches, look utilitarian. This does not shock us in as far as we primarily expect those sites to fulfill their function. Nevertheless, some authors see it as a major turning point in today's architectu-

5. See the magnificent pages written on New York City by C. Lévi-Strauss in *A World on the Wane*.

ral crisis and denounce the loss of what Christian Norberg-Schulz has called the *type* of public buildings, that is all the architectural characteristics that make us immediately distinguish between a government ministry, a church or a school, and a bank.[6] Between the technical prowess of the skyscraper and the often modest architecture of new public places, the very idea of monumentality seems to have been erased. How then can the majesty of public life still be expressed? Can it still claim any semblance of majesty? What remains of public life? How can it be expressed and realized if this is no longer the aim of the city? Must the city meet this demand or else lose its purpose as inhabited space?

The evolution toward undifferentiated buildings appears clearly *with the disappearance of urban enclosures.* When during the industrial revolution of the nineteenth century in Europe the old ramparts of many cities were destroyed, people believed that the perimeter had just been extended as had happened before in the past. All ancient cities experienced several stages of shifting enclosures. It seemed that the city had expanded and absorbed the immediate surroundings. In Paris for instance, there was a first circle, then a second, and a third. We thought that even stretched out like that, distorted, the city remained the city. In fact, it was the beginning of another history whose new direction we perceive today. The city was living its last days as a monumental city, as an organic whole. The city no longer has any limits. Strictly speaking, the city as such has disappeared. While the name remains, the idea of the city no longer exists, at least not as our former ideal. It remains to be seen if this erasure has left it with any life at all. But how can we consider the complexity, the size, the openness, and the articulation of urban volumes and especially the particular dignity of public buildings without acknowledging a debt to monumentality?

---

6. See C. Norberg-Schulz, *Intentions in Architecture* (Cambridge, MA: MIT Press, 1966). This lack of differentiation is not a new problem; at the end of the eighteenth century, the renowned architect C.-N. Ledoux believed it meant progress: "We will see magnificent open-air cafés and palaces of the same scale . . . Grandeur belongs to buildings of all kinds."

Surrounded by its walls the city was such a monumental and homogeneous whole that for a long time to enter the city meant to go through its doors which could be opened or closed like those of a house. For instance, we know that one evening when Jean-Jacques Rousseau was sixteen, he found the gates of Geneva closed, and decided to go into exile. The birth of the suburbs is but the prelude to the disappearance of the city and the appearance of a generalized urban space. In short, cities no longer have an outside; they no longer tend to contrast with the countryside. What we see are indeterminate spaces sprinkled with constructions. Agricultural spaces become what we call "green spaces," in other words, natural interstices between settlements. The most striking example today is the megapolis that stretches from Boston to Washington, or the immense conurbation from Tokyo to Kobe, or even Hiroshima. The urban archipelago sits on earth like a chain of islands on the ocean, without the beauty. The traditional countryside is not simply the nonurban. To speak of villages or farms is to evoke a millennial mode of living with their practices, their relation to the animal world, their knowledge, tastes, taboos, temporality, and wisdom. It means referring to a different form of civilization. To call it increasingly a "green space" is to admit that rural culture itself has disappeared or is disappearing. Today we may perhaps have only four kinds of non built-up spaces left, the wide open sea, high mountains, the desert, and the sky. Although we know what they are and have mapped them, on our over-urbanized planet they have become for us emblematic of our fascination with origins and common fictional references.

THE RETICULAR SPRAWLING OF MEGALOPOLISES

This brings us to another aspect of the crisis of traditional monumentality: beyond the overflowing out of the classic enclosures, the model of a scattered city is decisively established. The growth of ambitious contemporary megapolises, particularly in the developed world, forces us to face the following question; either this urban form represents a regrettable evolution, if not an irreversible failure, or we are dealing with the emergence of a new type of urban living, partly failing but perhaps in the process of succeeding or even emerging in a whole new way. A good example might be Los Angeles. Especially for visitors from an older civilization, the shocking aspect is above all *the disappearance of public space* in the most traditional sense of the term, that is, a space where the public sphere is expressed and represented as in the centralized architecture of the monumental city, organized around symbolic sites of authority and religion. But also lacking is the space *of the public* where a city offers a functional and pleasant daily life in what are called public or more accurately common spaces such as streets, sidewalks, shops, restaurants, squares, cafes, parks, museums, and theatres. Those two essential dimensions, public and common spaces, seem to have increasingly disappeared from cities in the United States. No one has said it more forcefully than Jane Jacobs in her famous book *The Death and Life of Great American Cities*.[7] She writes: "Los Angeles is an extreme example of a metropolis with little public life depending mainly on contacts of a more social nature. . . . Los Angeles is embarked on a strange experimentation: trying to run not just projects but a whole metropolis by dint of 'togetherness' or nothing. I think this is the inevitable outcome for cities whose people lack city public life in ordinary living and working."[8]

The collapse of public life in urban spaces has also been highlighted by R. Sennett.[9] This point has equally been emphasized by Mike Davis, in one of the few books, *City of Quartz*, dealing in depth if often hyperbolically, with the case of Los Angeles.[10] Is Los Angeles still a city? Or is it only as a negative, as we say in photography? Or does it perhaps foreshadow something else? Per-

---

7. J. Jacobs, *The Death and Life of Great American Cities* (New York: Vintage Books, 1961).
8. Ibid., p. 73.
9. R. Sennett, *The Fall of Public Man* (New York: Knopf, 1977).
10. M. Davis, *City of Quartz: Excavating the Future of Los Angeles* (New York: Verso, 1990).

haps one should discover it from an airplane, at night while descending toward the airport. Los Angeles then appears as a unity: an immense luminous rug like a starry sky fallen on earth. Perhaps this is the only way in which Los Angles looks like a monumental whole, like a city. During the day, driving from freeway to boulevard, from residential areas to avenues without pedestrians, the visitor wonders: but where is the city? "Los Angeles, writes Jean-Luc Nancy, is the prime example of a sprawling, scattered city that is extroverted, turned upside down, flattened out, and faces the sky belly up."[11]

Nevertheless there are encouraging elements in the case of Los Angeles, to counter Mike Davis' excessively pessimistic vision:

- *the concept of the city as an archipelago*: this is so-called *cluster* model as opposed to the *potato* model. The cluster model provides greater spatial availability and multiple centers. This is an essential point in a multicultural city because it gives each group the opportunity to find its own niche and local center in a global network, where no unique site embodies highly valuable space to the detriment of all others and where no ethnic neighborhood is a ghetto. The traditional, very centralized European model tends to identify minority groups with spatially peripheral sites and poor and neglected suburbs. The single-center model exacerbates the feeling of exclusion of the margins.
- *the concept of open circulation;* the circulation grid is neither radial nor concentric but consists of a system of indefinitely extensible rectangles. Such orthogonal grids are nothing new, but Los Angeles draws all the conclusions: public monuments like City Hall, the federal building, the LACMA museum, and the music auditorium built by Frank Gehry, are not convergence points but are set like other buildings along traffic axes and elude any perspective vision. And so we

11. J.-L. Nancy, *La ville au loin* (Paris: Mille et une nuits, 1999), p. 25.

could perhaps say that the true monuments of Los Angeles are its freeways.

- *the concept of a global urban space*; the enormity of Los Angeles' famously known urban sprawl obliges us to think of it as an *urban landscape* instead of a city. Perhaps the "monument" is the entire space; this is why at night it appears like a luminous galaxy crisscrossed by the incandescent rivers of the freeways. We should think of the site as an ecosystem, as do new urban planners.

We can compare this case with two other modern metropolises that have a longer history: Tokyo and Paris. Tokyo, created at the beginning of the Edo era, is not very old; its industrial development dates from the end of the nineteenth century during the Meiji reform. It is a significant case because public space has never been an essential feature in the Japanese tradition; nor did it have the centralized structure we have known in the West. Moreover, if one considers the various destructions, like the one that followed the earthquake in 1923 or the bombings in 1945 that led to the considerable renovation of older neighborhoods, one wonders if the reticular and decentered aspect of the city might be the result of an anarchic reconstruction without any urban planning (except in a few particular complexes). A completely different logic seems to govern this galaxy of neighborhoods with meandering streets, focused on local connections, as if social relations stubbornly prevailed over architectural intentions, if indeed there were any. Tokyo is a loose conglomeration of micro-cities that develop their own spirit, local life, relations of solidarity, and meeting places.[12] From that perspective, it is almost the total antithesis of Los Angeles.

The second case is Paris, the very exemplar of the ancient centralized city that has been typical of European urbanism for centuries. However, today's metropolitan Paris (more than 10 mil-

12. There is an important literature on Tokyo in Japanese, described by A. Berque in his book, *Japan: Cities and Social Bonds* (Yelvertoft Manor, Northamptonshire: Pilkington Press, 1997).

lion inhabitants) with its suburbs is almost comparable to Los Angeles in terms of size. It has its share of the same urban problems as other large metropolises: traffic, housing, role of supermarkets, archipelagos of housing projects, freeways, social problems, and criminality. Nevertheless, the permanent reference to a center, the Paris intra-muros, makes the rest of the city peripheral, while Los Angeles and Tokyo immediately became multicentered, for very different reasons. The off-centering of Parisian suburbs is inseparable from their feeling of having been abandoned. While that is not the whole problem, it remains its stormy horizon.

We must admit that ancient monumentality has lost its foundation and most of its bearings. As a result the concept of public space as visibility linked to monumentality is also in crisis. If this generalized urban phenomenon can still be called a city, a fragmented city, a city as an archipelago, it no longer cares about that visibility which remained a collective good for all to see, even in authoritarian political systems.

6

SECOND PROBLEM: THE CRISIS OF THE PUBLIC SPHERE

However, it seems that the concept of public space in the second sense cannot be ignored. The real question is whether the debate essential to democracy needs the type of physical space resembling the Greek and Roman model. In reality, since even the most open and free democracy cannot function without the mediation of duly elected and hence legitimate representatives, the "space" of public debate also presupposes intermediaries. Not all citizens need to be physically present at the site of the national assembly; opinions circulate in reviews, newspapers, and today in radio and audiovisual media, especially the Internet. In short, the space of public debate became a virtual space a very long time ago. Institutional debates that lead to voting on laws take place in appropriate sites and nowhere else, according to rigorous, publicly known and accepted, and controlled procedures. In short the form and validation of debates are themselves defined by law and this makes them different from debates that take place in the field of public opinion, including the most critical and best informed ones.

This leads us to consider the crisis situation linked to the concept of public space. It is not a physical "space." It evokes a formal model of space but actually functions solely in *systems of institu-*

tions and relations. We need to draw the conclusions implied by this fact.

CITY, PUBLIC SPACE, AND VIRTUAL SPACE

As we have seen, from the beginning the city has functioned as a network even if network theory itself is recent. This theory has taken on importance proportional to the changes in the information and communication techniques of the last decades. The specificity of noncentered networks is that they can be entered anywhere because, thanks to its multiple links, from every point (node) all others can be very quickly reached through successive connections. We deal with such systems in daily life when we use means of transportation or communication. Networks transform the relation between the local and the global, as Michel Serres has been saying consistently.[1] He tells us that the local is only a point of transit and relation because it is also either a node or an information center. As a result everyone is virtually everywhere. The ancient dream of ubiquity begins to take shape. Actually, it is realized in multiple ways. Bodies themselves can be ubiquitous as they change continents in a few hours in trips that in the past required weeks or even months. Even better, without going anywhere we can intervene in real time, simultaneously in different places, thanks to communication techniques, as in videoconferencing or group messages on the Internet.

We are dealing here with the virtual world; as such it is in no way an invention of our recent technologies. Language, the imagination, and fiction have always been ways of producing and engaging with virtual worlds. However, even confining ourselves to the question of so-called virtual space, we experience it constantly in everyday life. For instance, between two people who communicate by letter or telephone an invisible, unlocalizable space is

1. M. Serres, *Hermès IV, La distribution* (Paris: Minuit, 1977); *Hermès V, Le Passage du Nord-Ouest* (Paris: Minuit, 1980); *Atlas* (Paris: Bourin, 1994).

created that belongs to neither of them; this could also be said of a simple conversation. The space where our messages cross is halfway or even nowhere, elsewhere, or an empty point, rather than "in between" with respect to each site. This elsewhere is not added on to our experience but belongs to our human condition. Contemporary communication techniques therefore do not distort our spatial belonging or disturb a niche destined to remain local, because we have always inhabited the "elsewhere," or have been dislodged and shifted or distanced. They do not just extend our sense or motor organs. Still more, according to Serres, they actualize and *realize our representations*, the imagination in images, the voice in messages, and the elsewhere in networks of connected sites. Virtual space is not a simple space precisely because it contradicts the laws of the allocation of space and subverts the principle of the excluded third. I am always here and there at the same time; always going from the local to the global.[2] The city never ceases to teach us this belonging and to call for this uprooting.

Those are the terms in which we need to rethink the very notion of the "public sphere." When we say *space* or even *sphere*, we should be aware that we are dealing with a metaphor. What we designate as such is in fact a *network* of institutions and relations understood from a certain point of view, a specifically normative one concerning the responsibilities we recognize that we have toward one another in our dealings with these institutions. This is nothing new; it was already true in Greek cities even if because of their size the system was easier to recognize in the perceived and lived space of the city. What is new is that today's system has multiplied information and discussion means (both written and audiovisual) and therefore disseminated and inflated the old "virtual" at the time when urban space is losing its visibility as public space.

The crisis we experience today is similar to the one caused by the circulation of printed books at the turn of the fifteenth centu-

2. See M. Serres, *Atlas*.

ry. Today we must conceive of the public sphere—defined by Kant as citizens' public exercise of reason—based on the network model, with its multiplicity of sites and diversity of means of discussion and decision-making. This leads to the question whether the city itself has been rendered obsolete as public space in the most literal sense, that is, as architecture. Moreover, does this ruin its function as an information, decision-making, and cultural center? "Yes," are tempted to answer those both fascinated by and worried about the new powers of electronic networks. However, they may very well be wrong. What is more, we see the opposite tendency emerging. We must investigate this paradox.

What is the theoretical difficulty here? We can think about it as follows. From the time of the older model until its culmination during the Enlightenment, we have identified the concept of the public sphere with the notion of the *universal*. Perhaps this universality should not be rejected but conceived of differently. For a long time it was identified with the foundational model. It has often been noted that the underlying vocabulary of metaphysics has been architectural (basis, foundation, order, construction). Today however, common experience shows us that the universally understood geometrically homogenous space (isotropic and centered) appears as only one case or a limit inside a global space consisting of connected local multiplicities. In the classic definition, the universal escaped contingency. Since Aristotle we know formally: the universal is that for which there is a law that can be verified everywhere (the course of the planets for instance). The local, the very figure of the contingent, had no hope of acceding to the universal and hence to the truth of a law, unless it was recognized as having characteristics subsumable under the law. Only the universal could grant the local the dignity of being part of the whole. The local as such was outside of any possible salvation. As the singular, it remained the senseless, at most an exotic particularity. One had to choose: either be part of a totality defining it as legitimate or be excluded and thus occupy an unthinkable and absurd place: banished, outlawed, outside the walls; outside the public space of reason.

Global thought reverses the terms: the global does not exist before the local but it is the totality of their relations as a network. One can see the importance of such thinking to understand the modern cosmopolis. Already one truth is becoming more familiar and more friendly, that is, the contingency of our assignment in space. Why be here rather than elsewhere? This question may refer to the initial absurdity of the destiny that threw us here, somewhere on earth, or even generally to the fact of occupying any place at all. In the framework of a theory of the abstract universal, there is no good reason for such an assignment, which makes it contingent. This is why geography, the knowledge of contingency, could not acquire the dignity of a classical science since description is not explanation.

However, as soon as we leave the center/periphery model, as soon as the center is everywhere and the periphery nowhere, local settlement acquires a different status. Each point is an active center in the multiple intersections of the network. Every site is in real or virtual communication with all other sites. The virtual allows us to conceive of a form of universality both in principle and operationally because it simultaneously reaches every point of the network. In other words, it is a universality based not only on principle but also on connections. Each local point implies the global network; vice versa, the latter is nothing without multiple singular sites. We need to understand this in order to conceive of a new public space opening up for us beyond the old paradigm of architectonic universality, which invites us to think very differently about the city as we face its recent transformations.

THE RESISTANCE OF THE CITY AND THE TERRITORY IN NETWORK SPACE

Since the political stakes are clear, we will have to accept the new relation between the local and the global, which transforms our idea of public space. However, does this imply that we must assume that territories will inevitably disappear, or that cities will fragment into loose conglomerations to be managed by information and communication networks? If this hypothesis were correct, in recent decades we would have witnessed a strong regression of metropolises and megalopolises, especially in the developed world. However, the contrary has happened: demographic, economic, and cultural growth has been constant. Moreover, it is a major aspect of globalization.[3] It is tempting to see it as a complex but obscure aspect of the relation between humans and the built environment; for instance it could be a tendency to cluster together, according to the model of the so-called anthill. To counter predictions of the impending disappearance of the city, it would be more fruitful to think for a moment about the economic components of this phenomenon.[4]

Clearly, international companies tend to settle their services subsidiaries primarily in big cities while they locate their production activities in industrial parks connected to a metropolis. This shows the relevance of the rank/size Zipf indicator. This tendency is surprising given the high cost of office space and housing in prestigious cities, not to mention the cost and various inconveniences linked to transportation or pollution, for instance. Nevertheless, those disadvantages seem secondary compared to the advantages. What are they? Roughly speaking there are two sorts. The first are economic and technical. The large metropolis has important infrastructures in terms of office space, electronic connections, air transportation, and the important interaction of economic agents for negotiations, such as convention facilities and direct access to strategic information. Other advantages include easier contacts with large financial institutions and legal firms as well as capacious hotel and lodging accommodations. All this can result in economies of scale and critical gains in productivity.

3. Saskia Sassen, *Global City: New York. London, Tokyo* (Princeton: Princeton University Press, 1991); Pierre Veltz, *Mondialisation, villes et territoires, L'économie d'archipel* (Paris: PUF, 1996). Jean Rémy, *La ville phénomène économique* (Paris: Anthropos, 2000).

4. See Jean Rémy.

Significantly, at this level the proximity of other companies and fast relations among agents are understood to be direct economic advantages. It is even more instructive to consider what is specifically due to the urban setting in the second series of advantages, called external factors or noncommercial variables. Let us look at some of those "externalities," as economists call them. For instance, there is a concentration of competent agents and the proximity of educational institutions, such as universities, high-level institutes, and laboratories. There are also complex public administrative institutions including legal ones. More importantly, the urban milieu offers an environment where culture, knowledge, and communication stimulate research and innovation, something we have known for a long time; it has been referred to as a "permanent incubator," a telling formula. More generally, it offers the availability, diversity, and integration of resources called the agglomeration economy, which is no other than the city.

However, some scholars note that the entity "city" does not constitute a pertinent spatial category for economists who are interested in data and processes at the level of either groups of nations, single nations, or regions, but not cities except as poles within those territories. This attitude is changing with geo-economics. In that perspective one factor stands out as probably the most important: the *stability* of urban sites. Even if the city is a formidable transmitter and receiver of all sorts of flows (information, capital, energy, knowledge, agents, and goods), it is in a fixed position itself, a sort of immobile motor. This stability can be interpreted in purely functional terms. However, this would miss the essential point that the city is a *place*, and not a pure network settled on a *territory*[5] most often centuries ago; this spatiality, long manifested in monuments, is irreducible. It is also visible in the *density* influencing habitats, neighborhood relations, and professional activities. This spatial stability is incomprehensible without the inhabitants living in the space. The economic dynamics of

5. J. Lévy, *Le tournant géographique. Penser l'espace pour lire le monde* (Paris: Belin, 1999); Pierre Veltz, *Mondialisation, villes et territoires. L'économie d'archipel.*

urban poles cannot be understood without taking into account all the elements comprising the city.

To be sure, there are the most obvious "externalities," but there are many other things not even considered by economists, for this would take the economy too far outside its field of analysis.* These elements include everything that keeps the city together as an assembly of humans; undoubtedly it works, produces, and exchanges, but it exists because it is the heir of a memory and remains the terrain of social and political struggles. The city is legally defined by institutions and regulations but also lives from day to day, and is made up of individuals and groups connected in various ways, and who love, or don't love, the place where they live, have relationships, activities, routines, and their destinies. This is the space they are given, and given in common as the very word of community says literally. We must understand it.

*Because it is an illegitimate field that has no understanding of its true object, humans

7

THIRD PROBLEM: REDISCOVERING COMMON SPACE

We accepted long ago the satisfying canonical opposition between *private* and *public*. We must hasten to add however, that it belongs more to the discipline of political philosophy than of sociology and anthropology. Its great heuristic power makes it particularly useful to describe two fundamental modes of human life. The theoretical depth of the work of Hannah Arendt and Jürgen Habermas did much to provide the model for the opposition between private and public. Nevertheless, those two terms no longer suffice to cover the subject. They force us to extend the private sphere outside the home or to indiscriminately push into the public sphere anything that has social visibility. We need another concept. The label is fairly obvious: *common space*. The immediate objection is that such a concept already exists and could also correspond to what has been designated as the entire social domain. It is doubtful however that we are dealing with the same thing. Arendt's analyses are instructive. She uses the term *common* to designate foremost the content of the public domain.[1] She argues that what we have in common is the business of the city. Citizens

1. H. Arendt, *The Human Condition* (1958) chapter II, third section: "The Public Realm: The Common."

taking care of the city's business is the action, *praxis*, concerning collective decisions, law-giving, in short, public life. What then becomes of the *social*? This is a fairly recent concept but can be retroactively applied to older societies. For Arendt it concerns everything that has to do with labor activities: relations of production, exchange, and consumption in the city. It is the universe of *homo laborans*.

We have that in common as well; it must be visible to others and obtain the status of *appearing before all*, which pulls it away from its particular character and from merely useful existence. The artisan, for instance, who shows his wares in the market, not only wants to sell them but also shows them to have them acknowledged as the result of his labor. This is a profound statement. In promoting the common, Arendt consistently favors the optical model: according to her, what is *visible* to all is what is recognized as public. However, the social world remains defined in very general features. As a result, these features can be applied to various cultures and periods. Without denying the pertinence of the social as now generally understood, in the context of the urban phenomenon the *common world* needs to be approached from a different perspective. Below are outlined the main features in order of increasing importance and inclusiveness.

The common world is made up of interrelational practices such as neighborhood relations and everything that expresses a shared and familiar lifestyle marked by either random encounters as for instance the comings and goings in the streets, shopping, visits to inns and cafés, and the use of public transportation; or the more organized encounters of association and brotherhood meetings, civil or religious ceremonies, neighborhood or workplace celebrations, sporting events, and musical parades. More fundamentally, the common world overlapping between public and private comprises mores and civilities, and, more generally, traditions of all kinds: food and clothing choices, gestures expressing emotion, language usage, religious and religiously inspired attitudes, contacts between men and women, between children and adults, between members of the same profession, workplace relations, or reactions to status differences.

In short, we are dealing with the *vernacular order* of ways of doing and saying that mark the local and configure uniqueness, in other words what we mean by *styles*. Those practices of everyday life have always constituted the substance of city life. They confer color and particularity on everything we mean by social life. They create the *atmosphere* within which the relationship to public space is experienced and understood. We could say that the last circle is drawn by the civilization to which a given city belongs. This is undeniably true but we have to qualify this by noting that each cultural tradition, each city, and each neighborhood tends to develop a unique common world, often for a fleeting period. What remains of Shanghai of the thirties for instance; or Proust's Faubourg Saint-Germain; Man Ray's Montparnasse; Gold Rush San Francisco; Pessoa's Lisbon? Perhaps the common world falls within the province of a meteorology of atmospheres that are specific to groups or of a geography of tastes and emotions linked to sites, which should be described by some future urban ethology. If so, the urban space that would be its proper topic is the street, and in a different way, the square of modest dimensions as well.

THE STREET, THE SQUARE: SITES OF COMMON SPACE

The street is not a monument: strictly speaking it is only a void between buildings and space for pedestrians and vehicles to circulate. However, in many respects it condenses the reality of the city; it reveals its atmosphere, style, rhythm, charm, surprises, and sometimes its defects and obstacles. To understand the street is perhaps to understand the reason for the existence of the urban phenomenon and the longing for the city. This is why the street as the first necessary form of the urban network raises the ultimate question about built space at a time when all sorts of networks triumph. It is the space par excellence for the circulation of bodies which forces us to face the simple fact that the city exists for those who inhabit it. Evidently, many architects, urban planners, and political authorities need to remember this. Ostensibly, it was the main concern of the adherents of Modernism. Ironically, however, they were the ones who, more than any other professionals in the business, displayed a kind of contempt for anything in the urban space that might have a local context, particular values, or a vernacular order. This could be the last avatar of the increasing importance of the representation of homogeneous and undifferentiated space that was so successful in Galilean physics, ballistics, and the construction of fortified towns. However, it was definitely not suitable for urban planning or at best could be used for building railroad stations, warehouses, and suchlike, but not living spaces. Le Corbusier's anathema against the street is well-known: he kept calling it the "donkey path" and wanted to make it disappear forever, especially if it curved.[2]

Behind the problem of the street looms the unsolved problem of contemporary urban planning: how to structure the growth of the network-city that must meet the demand for a comfortable life in terms of transportation, access to services and communication, and health requirements such as light, good air, greenery, and spaces for leisure activities. It should also favor professional efficiency with adequate offices and fast connections; moreover it would have to take into account the desire to live together in a densely populated built environment without overpopulation and to offer security without ending up in overprotected encampments such as the gated communities of the United States.

But we should not idealize one component, "the Street," as a stable and calibrated form. We might even need a typology distinguishing streets according to size: an alley for instance, a simple passage between blind walls, cannot give you the sense and atti-

2. Le Corbusier writes: "The corridor-street with two sidewalks, suffocating between tall houses, must disappear. . . . A modern city lives with the straight line. The straight line is also healthy for the city's soul. The curve is expensive, difficult and dangerous and it paralyses. The straight line is everywhere in history, in all human intentions and actions . . . the curving street is a donkey path; the straight street the human path." Le Corbusier, *Urbanism* (Bombay: J. J. Bhabha, [1925] 1952–1953).

tude of the "street" type. Or it could be according to the height of the buildings: even a very active street in Manhattan[3] is not like a street in Naples or Kyoto. Then we would need to distinguish between streets with or without businesses, cafés, or restaurants: a street without them is reduced to a dreary thoroughfare. Next we need to establish what contribution is made by the street's relations to cross streets, or to the squares and avenues it flows into. Moreover, the quality of the buildings and their age would also have to be taken into account. Finally, we must not give in to the temptation of referring primarily to streets in western cities. The legacy of urban forms in China, India, Japan, the Middle East, and other forms of outdoor encounters do not grant the street quite the same function or status, even if modern buildings tend to create the same types of urban spaces everywhere.

Nevertheless, let us agree that in speaking of the street here, we will focus on the normal street of western or similar cities with sidewalks, shops, restaurants, and most often cars, buses, or trams. Why give the street so much importance? Probably because it exemplifies the most specifically urban sociality which from the very beginning reinvented itself by breaking away from what were earlier lifestyles, essentially those of villages or nomadic encampments.

The street is the site par excellence of what could be called *common life* in that it is distinct from both private and public life. The street gives shape to this common life in terms of at least four components: vicinality, civility, visibility, and diversity.

Vicinality. The inhabitants of a street tend to feel a certain solidarity simply by the feeling of belonging to the same place. Most often this remains implicit, especially when there are no opportunities to meet or talk; in time neighbors will recognize each other after they keep meeting in the same shops or on the same streets; this feeling is experienced more intensely in small towns. This neighborhood relation is actually quite strange: it is not a kinship relationship, nor one we might have in public spaces.

3. For Jane Jacobs, New York City streets are identified with wide sidewalks.

It is simply shared space, comparable to the space provided by a small square but less obvious to those who live on an avenue or boulevard. Neighbors have this peculiar thing in common: they are not chosen as we choose our friends but they are nevertheless close to us simply by belonging to the same site. At the same time, this feeling of belonging remains at a kind of respectful distance only overcome by those who regularly go to the same bar or club. The street gathers and permanently tests those familiar and reserved neighboring relationships that are at the heart of urban living. Those seemingly unimportant informal relations contain a paradox so well stated by Jane Jacobs: "Most of it [the contacts] is ostensibly utterly trivial but the sum is not trivial at all."[4] This sum, she says is, a feeling of trust circulating throughout urban space.

Civility. The city is also par excellence the site where people come and go. Because they constantly run into their neighbors, they recognize them and show this either by verbal courtesies if they are close enough, or by looks and gestures full of benevolence; at least they make efforts not to offend when passing by, to maintain a polite reserve. Erving Goffman[5] gives an admirable analysis of the attitudes and rituals arising from those constant informal contacts in street spaces. This street civility is all the more remarkable because of the assumption that those people encountered for a brief moment will probably forever remain unknown to us. How can this enigmatic generosity be understood? One could say that common space is experienced first as a peace offering and an occasion for serenity: it is neither private space to

4. J. Jacobs, *The Death and Life of Great American Cities*, p. 56.

5. E. Goffman, *Encounters: Two Studies in the Sociology of Interaction* (Indianapolis: Bobbs-Merrill, 1961); *Interaction Ritual: Essays on Face-to-Face Behavior* (Garden City, NY: Doubleday, 1967); *The Presentation of Self in Everyday Life* (Garden City, NY: Doubleday, 1959). See also P. Dumouchel, *Tableaux de Kyoto* (Québec: Presses de l'Université Laval, 2004), p. 30–35. Dumouchel shows that what is at stake in public and private is not the same in Japan and the West. In Japanese urban space the gaze does not function as a means to a dialogue or a duel. Only in the public space of *institutions* is one obliged to save *face*, not in the common space of the street that is defined as being shared individually. It follows that for Japanese sociologists, the analyses of E. Goffman, apart from their methodological interest, are valuable as ethnographic documents about New York, but not very relevant for their own societies.

be defended against intruders nor public space ruled by overly strict protocols. It is space that belongs to *us*, as members of a varied humanity manifesting or inventing itself in unexpected circumstances, conscious of our responsibilities. We enjoy this burden as an exercise of our freedom. We appreciate one another as so many embodiments of *Everyman*. The street makes us available for things that are nowhere programmed. This is undoubtedly the pleasure felt by the *flâneur* according to Baudelaire and, reading him, Walter Benjamin. In this context we emphasize the importance of the *sidewalk*, the reassuring margin separating the pedestrians from the gutter and protecting them from the stream of cars. It draws them toward the space of shops and cafés; above all, by concentrating the continuous stream of pedestrians, the sidewalk incessantly provokes the brief encounters renewed by the to-and-fro of bodies, as faces advance and glances meet and sparkle.

Visibility. The street is generally a site where people pass. Functionally, it is a simple space for circulation to go from one point to another. It is perhaps especially a site for encounters. However, those are strange encounters consisting primarily in ephemeral crossings of people unknown to each other. Where does this strange imperious desire to stroll in the streets come from? One could answer that it might be the exceptional opportunity to show yourself: to be seen without being recognized, at least away from one's own street. This is why we do not go out into the street without some attention to our physical appearance or some desire to please or at least not to displease. In a way we can show ourselves without cost or demands because there is nothing at stake. This brings about a light, a limited theatricality that precisely distinguishes common space from public space, as a stage for a nonstrategic form of visibility that involves seduction and play rather than power or duty. The other's response is tested in their attention or failure to look.[6] It combines the happiness of being oneself while being nobody. The street offers everyone without exception the opportunity to see and be seen; to be there simply among others; to be unique in a banal environment; to be welcomed without being named; to be with others without commitment to anyone. However, the concern for appreciation and for receiving respect *for one's appearance* is evidence of a truth: in the common space *all the others are important to us*. It is on the basis of the sociability of this neutral world that more explicitly shared common life becomes possible and remains open to all, the life of meetings, festivals, discussions, associations, and responsibilities.

Diversity. The street is probably the only urban space where all individuals in a society have a chance to pass one another, whatever their social background, age, gender, profession, ethnic origin, or belief. The street is the melting pot of the greatest total diversity. Even bars, shops, official buildings, department stores, or places of worship can only offer this partially since there will always be some group in the population that never goes there. Moreover, other sites that are legally accessible to all, such as post offices, hospitals, schools, and sports centers, are part of a more formal public space. The street is the incomparable space where diversity is expressed and welcomed given the variety of shops, offices, and restaurants where the clientele is heterogeneous. It remains so because it has no program and cannot propose any. By definition the street is unpredictable. It becomes unique, known by name, because its singularity emerges from the improbable mixture of the thousand differences it contains. This is why the street, barely separated from private space by doorsteps, represents common space, and can also change into public space, with political or union marches, religious or sports ceremonies. This is not surprising because the street is above all a passage for everybody. What this means becomes very clear when intolerance,

---

6. This concerns primarily the dominant attitude in the West. In Asia, and particularly in Japan, there is an art of seeing others without looking at them, which is suitable for common space and changes in private and public space. See preceding note.

whether racist, religious, or social, refuses access to this or that street to certain city dwellers. Such prohibitions are intolerable since they amount to negating the city. More fundamentally, the street is experienced as an emblem of freedom, as unconditional open space, and finally as an expression of democracy. Precisely when it is threatened, we know what "to let the street speak" means.

NOTE ON THE SQUARE

These remarks on the street also concern the square, at least the small one that thanks to its cafés or shops is a much sought-after site of common life. Of course it can be crossed like a street, but it offers something quite different. There is something gratuitous about its existence. It offers an opening through the mass of the buildings and invites one to look more, to raise one's eyes and to linger. The street always goes somewhere, while the square remains in place. Often automobile traffic only reaches one side or does not even enter it. The buildings on each side look at each other and seem to be having a meeting. They invite us to do the same. It is like an open-air living room. Of course, the question of size intervenes: if the square becomes too large, like Red Square, Place de la Concorde, or Tiananmen Square, you do not get that feeling but they are better suited for public ceremonies. In general, the small square cannot be seen from afar; it is around the corner or rather it appears suddenly by surprise and becomes a resting place for the street. At night, when the cafés are closed, the city seems to collect itself there. The street and the square have the power to give a rhythm to urban space and invite us to adopt its breathing.

CONCLUDING REMARKS: THE FUTURE CITY

DWELLING IN THE CITY, DWELLING IN THE WORLD

We need to return to our earlier amazement and ask again what could have impelled, more than ten millennia ago, human groups to build cities; to create dwellings that were more densely populated than villages but also included official buildings giving those built ensembles a monumental character; in short, to build cities? This achievement presupposed, exploited, and sometimes enslaved labor and established hierarchies that were probably unknown in preceding times. Was it really such a good thing? We know the essential role played by the emergence of agriculture in this transformation. Could humanity have invented other solutions and created another way of living together? How can we understand the very act of building? We could like Heidegger assume a fundamental articulation between building, living, and thinking. However, we cannot, as Heidegger does, make the sole reference to building the isolated peasant dwelling at the edge of the forest and discuss the city only in the context of a housing crisis. The relation between the three terms, building, living, and thinking,[1] should be explored more thoroughly when dealing with the emergence of cities. They very much concern our lives on earth and our way of being in the world. By its monumentality, however, the city manifests a particular relation of buildings to the glory and the grandeur of inhabiting them. This requirement is probably just as inseparable from the longing for the city as is the longing caused by the experience of the close proximity of dwellings. Living in the city then is also living in the world but very different from living in a house on the hillside. It means living in the world because the work of construction that assembles a human community also signifies the gathering and welcoming of the world in the city walls: heaven and earth, the gods and humans. All civilizations thought the same thing. Monuments were to celebrate this dignity and architecture's task was to affirm this over and over again. Today that task remains. No network logic will make it disappear. When referring to Hugo's famous text: "This will kill that. The book will kill the building"[2] one generally forgets to mention the poet's other assertion: "The cities of men need monuments, otherwise, what would be the difference between a city and a[n] ant-hill?"[3] With all due respect for ants, the constant creativity of contemporary architecture confirms that the requirement of monuments remains at the heart of the city we are awaiting. What will it be? For the last time, let us address the question our epoch faces.

THE SITE OF SHARED LIVING

We asked at the beginning: are we facing the end of the city now that the urban form as the dominant housing mode is spreading

---

1. Martin Heidegger, "Building, Dwelling and Thinking," [1951] in *Poetry, Language, Thought,* translated by Albert Hofstadter (New York: Harper Colophon Books, [1951] 1971).
2. Victor Hugo, *The Hunchback of Notre Dame* (New York: E. P. Dutton & Co. Inc., 1958), IV–II, 164 sq.
3. V. Hugo, *Littérature et philosophie mêlées* (Paris: Robert Laffont [1834]; coll. Bouquins, 1985), p. 130.

over the entire planet? Rather than cities, we found we were dealing with an increasing global archipelago of *inhabited zones* or *urban modules*. They no longer seem to need privileged architectural sites that are centered, imposing, and visible to all. Should we acknowledge and lament the failure of our ancient urban civilization? Is it not more urgent to understand that the old purposes of the traditional city have disappeared? Does the expansion of the network model and remote communication that gives us new opportunities mean that we have to accept the *community of absent bodies*? An analysis of common space shows that the answer is clearly no. Hence it is more important than ever to rethink urban planning and architecture. We need to ask again what meaning to give to *the built environment* in order to articulate both sensibly and *sensorily* the social functions and architectural forms of the contemporary city. It is now a space less and less dependent on the immediate presence of bodies, where bodies nevertheless exist and feel no less than in the ancient Sumerian, Greek, Inca, Indian, or Roman cities but must express this space differently. How can we conceive of the built environment in such a way as to make communal life possible, as public space in the information age? This is the challenge.

In the context of virtual space how can we maintain concrete spaces for bodies, living space, and space for institutional and personal relations? How can we reinvent the *street* and the *square* and the pleasure of being together? How can we have *public*, that is *visible*, signs of an urban identity, and at the same time understand that the new communication technologies redefining our access to information, our sharing of knowledge, and our modes of action also radically modify our relation to the physical world, other humans, and other cultures? Ultimately they change our global perception of space and time, as well as our sense of communal life. To perceive the movement going on before our eyes from the *monumental* to the *virtual* in the coming city would be to understand one of the major transformations of our time. More importantly, we need to learn or relearn the movement that goes from the *virtual* to the *physical*. The synthetic image will never abolish the body. Network relations will never abolish the words that I say to the being that is closest to me or even to just anybody. We should understand that all this is not taking place in face to face encounters of purely abstract subjects. It is played out in the *place* where I speak a language, where I hear several languages, where I come and go, where I find my bearings, where I inherit the art of living and discover new ones, where I am a man or woman, where I work or look for work. It is also the place where I belong to a generation, where I talk occasionally to my neighbors, where my profession or my family and my origins are known, where parents form original relations through their children. Every city dweller is a bouquet of singular experiences in the field of common life. This is the niche of eco-urban relations wherever the city has retained neighborhood life, wherever it is still a city and remains a *livable* city, and gives us the certainty that it indeed constitutes our residence on earth. Utopia does not consist of an architecture that would clearly delineate spaces where all useful functions would be smoothly assured. Utopia is what used to be our daily reality in the West and elsewhere: a place where people talk to each other, look at each other, respect one another and quarrel, help out if necessary, where one can meet without needing a watch or a calendar. This utopia really exists since it remains the way of living of billions of people; this is the life we can almost no longer imagine in developed metropolises.

We knew it all along: "There are no more islands."[4] Some will say there is only one left, the planet itself. But from the proper distance of neighborhoods and encounters, daily itineraries and customs, what we see and experience is an archipelago of sites we insist on calling a city. We want the name to remain. In some places the city resists and exists because we appreciate not only all those things that meet our needs (businesses, services, leisure, and transportation) but all that makes us happy to live and work there, to interact, walk, see friends, feel other presences, and encounter

4. Albert Camus, "The Minotaur," in *The Myth of Sisyphus and Other Essays* (New York: Vintage Books, 1959).

unknown faces. In such cities we can sometimes see in the quality of the architecture the realization of the idea of the city that has survived all the metamorphoses. It is a common good that is precious to us in an entirely selfless way, a space constructed for all, for our senses and eyes, because the requirement and the pleasure felt in being a city dweller also means being a citizen. Beyond even this satisfaction, the city *in its visible form*, in its spatial organization and the intelligence of its design, can make us proud of living together, just as did the most ancient cities, and preserve intact within us the hope that the site of our shared lives can be admirable and desirable.

BIBLIOGRAPHY

Arendt, Hannah. *The Human Condition.* Chicago: University of Chicago Press, 1958.

Aurenche, O. *La Maison orientale: L'architecture du Proche Orient ancien des origines au milieu du quatrième millénaire.* Paris: Ed. Geuthner, 1981.

Balazs, Étienne. "Les villes chinoises: Histoire des institutions administratives et judiciaires." In *La Bureaucratie céleste.* Paris: Gallimard, 1968.

Benevolo, Leonardo. *The European City.* Cambridge, MA: Blackwell, 1993.

Benveniste, Émile. *Problems in General Linguistics.* Translated by Mary Elizabeth Meek. Coral Gables, Florida: University of Miami, 1971.

Berque, Augustin. *Japan: Cities and Social Bonds.* Yelvertoft Manor, Northamptonshire: Pilkington Press, 1997.

Billeter, Jean François. *La Chine trois fois muette.* Paris: Allia, 2006.

Blumenberg, Hans. *The Legitimacy of the Modern Age.* Translated by Robert M. Wallace. Cambridge, MA: MIT Press, 1983.

Boyer, Marie-Christine. *CyberCities. Visual Perception in the Age of Electronic Communication.* New York: Princeton Architectural Press, 1996.

Braudel, Fernand. *Capitalism and Material Life, 1400–1800.* Baltimore: Johns Hopkins University Press, 1977.

———. *Civilization and Capitalism 15th–18th Century.* New York: Harper & Row, 1982–1984.

Braunfels, Wolfgang. *Abenländische Stadtbaukunst.* Köln: Dumont, 1976.

———. *Mittelalteriche Stadtbaukunst in der Toskana.* Berlin: G. Mann, 1953.

Castells, Manuel. *The Rise of the Network Society.* Cambridge, MA: Blackwell, 2 vol., 1996–2000.

Cauvin, Jacques. *Les premiers villages de Syrie-Palestine du IXe au VIIe millénaire avant Jésus-Christ.* Lyon: Maison de l'Orient, 1978.

———. *The Birth of the Gods and the Origins of Agriculture.* New York: Cambridge University Press, 2000.

Cerdá, Ildefons. *Teoría general de la urbanización.* Madrid, 1868–1871.

Chandler, Tertius. *Four Thousand of Years of Urban Growth.* New York: Academic Press, 1974.

Charre, Alain. *Art et urbanisme.* Paris: PUF, 1983.

Choay, Françoise. *The Rule and the Model: On the Theory of Architecture and Urbanism.* Translated by Denise Bratton. Cambridge, MA: MIT Press, 1997.

Clavel, Monique, and Pierre Lévêque. *Villes et structures urbaines dans l'Occident romain*. Paris: A. Colin, 1971.
Cohen, Jean-Louis, and Hubert Damisch. [Dir.] *Américanisme et modernité. L'idéal américain dans l'architecture*. Paris: EHESS/Flammarion, 1993.
Damisch, Hubert. *The Origin of Perspective*. Cambridge, MA: MIT Press, 1994.
Davis, Kingsley. *Cities: Their Origins, Growth and Human Impact*. San Francisco: W. H. Freeman, 1973.
Davis, Mike. *City of Quartz: Excavating the Future in Los Angeles*. New York: Verso, 1990.
Descola, Philippe. *Beyond Nature and Culture*. Chicago: University of Chicago Press, 2013.
Detienne, Marcel. *The Masters of Truth in Archaic Greece*. New York: Zone Books, 1996.
———. [Ed.] *Tracés de fondation*. Paris-Louvain: Peeters, 1990.
Dreier, Peter, John H. Mollenkopf, and Todd Swanstrom. *Place Matters: Metropolitics for the Twenty-first Century*. Lawrence, KS: University Press of Kansas, 2004.
Duby, Georges. [Dir.] *Histoire de la France urbaine*. Paris: Seuil, 1979.
Dumouchel, Paul. *Tableaux de Kyoto*. Québec: Presses de l'Université Laval, 2005.
Dupuy, Gabriel. *L'urbanisme des réseaux: théories et méthodes*. Paris: A. Colin, 1991.
Fiévé, Nicolas. *L'architecture et la ville du Japon ancien. Espace architectural de la ville de Kyôto et des résidences shôgunales aux 14e et 15e siècles*. Paris: Maisonneuve et Larousse, 1996.
Finley, Moses. *The Ancient Economy*. Berkeley: University of California Press, 1973.
———. *Early Greece: The Bronze and Archaic Ages*. London: Chatto & Windus, 1970.
Geddes, Patrick. *Cities in Evolution*. London: William & Norgate, 1915.
Giedion, Sigfried. *Space, Time, and Architecture*. Cambridge, MA: Harvard University Press, 1962.
Ginzburg, C. *The Night Battles: Witchcraft and Agrarian Cults in the Sixteenth and Seventeenth Centuries*. Baltimore: Johns Hopkins University Press, 1983.
Glassner, J. J. *La Mésopotamie*. Paris: Belles Lettres, 2002.
———. *Mesopotamian Chronicles*. Leiden & Boston: Brill, 2005.
Grandazzi, Alexandre. *The Foundation of Rome: Myth and History*. Translated by Jane Marie Todd. Ithaca, NY: Cornell University Press, 1997.
Granet, Marcel. *Chinese Civilization*. New York: Meridian Books, 1958 [1930].
Grimal, Pierre. *Roman Cities*. Translated and edited by G. Michael Woloch. Madison: University of Wisconsin Press, 1983.
Habermas, Jürgen. *The Structural Transformation of the Public Sphere*. Cambridge, MA: MIT, 1989 [Orig. 1962].
Harvey, David. *The Urbanization of Capital*. Oxford: Basil Blackwell, 1985.
Havelock, Eric. *Origins of Western Literacy*. Toronto: Ontario Institute for Studies in Education, 1974.
Heidegger, Martin. "Building, Dwelling and Thinking." In *Poetry, Language, Thought*. Translated by Albert Hofstadter. New York: Harper Colophon Books, [1951] 1971.
Hénaff, Marcel. *Le lieu du penseur*. Paris: Cahiers de Fontenay, 1983.
———. "The Cannibalistic City: Rousseau, the Large Number and the Abuse of the Social Bond." In *Substance* 63 (1992).
———. "The Stage of Power." In *SubStance* 80 (fall 1996): 16–21. "Politics on Stage: The Machiavellian Moment and the Birth of Perspective."
Hénaff, Marcel, and Tracy B. Strong. *Public Space and Democracy*. Minneapolis: University of Minnesota Press, 2001.

Holston, James. *The Modernist City: An Anthropological Critique of Brasilia.* Chicago: University of Chicago Press, 1989.
Homo, Léo. *Rome impériale et urbanisme dans l'Antiquité.* Paris: A. Michel, [1951] 1971.
Huot, J. L. [Ed.] *Préhistoire de la Mésopotamie.* Paris: Ed. du CNRS, 1987.
Jacobs, Allan, and Donald Appleyard. "Toward an Urban Design Manifesto." *Journal of the American Planning Association* 53, no. 1 (1987).
Jacobs, Jane. *The Death and Life of Great American Cities.* New York: Random House, 1961; reprinted at New York: Vintage Books, 1992.
Judge, David, Gerry Stoker, and Harold Wolman. *Theories of Urban Politics.* Thousand Oaks, CA: Sage Publications, 1995.
Koolhaas, Rem. *Delirious New York: A Retroactive Manifesto of Manhattan.* London: Academy Editions, 1978.
Kostof, Spiro. *City Shaped: Urban Patterns and Meanings through History.* London: Thames and Hudson, 1991.
Kotkin, Joel. *The City: A Global History.* New York: Random House, 2006.
Kramer, S. N. *History Begins at Sumer.* Garden City, NY: Doubleday, 1959.
———. *The Sumerians: Their History, Culture, and Character.* Chicago: University of Chicago Press, 1963.
Lachmanni, Carl. *Gromatici Veteres.* Berlin: [Rééd, 1848–1852] Bari, 1960.
Le Corbusier, *Urbanism.* Bombay: J. J. Bhabha, [1925] 1952–1953.
Le Goff, Jacques. *Intellectuals in the Middle Ages.* Translated by Teresa Lavender Fagan. Cambridge, MA: Blackwell, 1993.
———. *Medieval Civilization, 400–1500.* Translated by Julia Barrow. Oxford: Basil Blackwell, 1988.
———. *Time, Work, and Culture in the Middle Ages.* Translated by Arthur Goldhammer. Chicago: Chicago University Press, 1980.
Lévêque, P., and P. Vidal-Naquet. *Cleisthenes the Athenian: An Essay on the Representation of Space and Time in Greek Political Thought from the End of the Sixth Century to the Death of Plato.* Atlantic Highlands, NJ: Humanities Press, 1996.
Low, Setha M. *Theorizing the City: The New Urban Anthropology Reader.* New Brunswick, NJ: Rutgers University Press, 1999.
Lynch, Kevin. *The Image of the City.* Cambridge, MA: MIT Press, 1960.
Malamoud, Charles. "Sans lieu ni date. Note sur l'absence de fondation dans l'Inde védique." In Marcel Detienne [ed.], *Tracés de fondation.* Paris-Louvain: Peeters, 1990.
Margueron, Jean-Claude. *Les Mésopotamiens.* Paris: A. Colin, 1991.
Martin, Roland. *L'urbanisme dans la Grèce Antique.* Paris: Picard, 1956.
Marvin, Simon, and Stephen Graham. *Splintering Urbanism: Networked Infrastructures, Technological Mobilities and Urban Condition.* London: Routledge, 2001.
McAdams, R. C. *Heartland of Cities.* Chicago: University of Chicago Press, 1981.
Mitchell, William J. *City of Bits: Space, Place, and the Infobahn.* Cambridge, MA: MIT Press, 1995.
Mongin, Olivier. *La Condition urbaine.* Paris: Seuil, 2005.
———. *La Ville des flux.* Paris: Fayard, 2013.
Mumford, Lewis. *The City in History: Its Origins, Its Transformations, and Its Prospects.* New York: Harcourt, Brace & World, 1961.
Nancy, Jean-Luc. *La Ville au loin.* Paris: Mille et une nuits, 1999.
Nissen, H.-J., and J. Renger. (Eds.) *Mesopotamien und seine Nachbarn.* Berlin: D. Reimer, 1982.
Norberg-Schulz, Christian. *Architecture: Meaning and Place.* New York: Electa/Rizzoli, 1986.

———. *Genius Loci: Towards a Phenomenology of Architecture.* London: Academy, 1979.
———. *Intentions in Architecture.* Cambridge, MA: MIT Press, 1966.
Panofsky, Erwin. *Gothic Architecture and Scholasticism.* Latrobe, PA: Archabbey Press, 1951.
———. *Perspective as Symbolic Form*, MIT Press, 1991.
Park, Robert. Ed. *The City.* Chicago: University of Chicago Press, 1925.
Parrot, André. *Sumer: The Dawn of Art.* [Paris: Gallimard, 1960] New York: Golden Press, 1961.
Payot, Daniel. *Le Philosophe et l'architecte.* Paris: Ed. Payot, 1983.
Pérez-Gómez, Alberto. *Architecture and the Crisis of Modern Science.* Cambridge, MA: MIT Press, 1983.
Pinchemel, Philippe, et Geneviève Pinchemel. *La Face de la terre.* Paris: Armand Colin, 1988.
Pirenne, Henri. *Medieval Cities: Their Origins and the Revival of Trade.* Garden City, NY: Doubleday, 1956 [1925].
Ragon, Michel. *Histoire de l'architecture et de l'urbanisme.* Paris: Seuil, 1991.
Reader, John. *Cities.* New York: Grove Press, 2004.
Rossi, Aldo. *The Architecture of the City.* Cambridge, MA: MIT Press, 1982.
Sainte-Lagüe, André. *Les Réseaux (ou graphes).* Paris: Gauthier-Villars, 1926.
Sassen, Saskia. *Cities in a World Economy.* Thousand Oaks, CA: Pine Forge Press, 1994.
———. *Global City: New York, London, Tokyo.* Princeton: Princeton University Press, 1991.
Savage, Mike, and Alan Warde. *Urban Sociology, Capitalism and Modernity.* Basingstoke: Palgrave Macmillan, 1993.
Schabert, Tilo. *Die Architektur der Welt: Eine kosmologische Lektüre architektonischer Formen.* München: Fink, 1997.
Sennett, Richard. *The Fall of the Public Man.* New York: Knopf, 1977.
———. *Flesh and Stone: The Body and the City in Western Civilization.* New York: W. W. Norton, 1994.
Serres, Michel. *Atlas.* Paris: Bourin, 1993.
———. *Hermès IV, La distribution.* Paris: Minuit, 1977.
———. *Hermès V, Le Passage du Nord-Ouest.* Paris: Minuit, 1980.
———. *The System of Leibniz.* Manchester, UK: Clinamen, 2001 [1968].
Sfez, Gérald. Machiavel, *La politique du moindre mal.* Paris: PUF, 1999.
Sitte, Camillo. *Städtebau nach einem künstlerischen Grundsätzen* [1889]. Translation. *City Planning According to Artistic Principles.* London: Phaidon Press, 1965.
Tonkiss, Fran. *Space, the City and Social Theory: Social Relations and Urban Forms.* Cambridge, MA: Polity Press, 2005.
Unwin, Raymond. *Town Planning in Practice.* Princeton, NJ: Princeton University Press, [1909] 1993.
Vernant, Jean-Pierre. *Myth and Thought among the Greeks.* London: Routledge & Kegan Paul, 1983.
Vidler, Anthony. *The Writing of the Walls: Architectural Theory in the Late Enlightenment.* Princeton: Princeton Architectural Press, 1987.
Virilio, Paul. *Speed and Politics.* Los Angeles: Semiotext(e), 2006.
Vitruvius: *The Ten Books on Architecture.* Herbert Langford Warren (Illustrator), Morris Hickey Morgan (Translator). New York: Dover Publications, 1960.
Wallerstein, Immanuel. *The Capitalist World-Economy.* Cambridge: Cambridge University Press, 1979.
Weber, Max. *The City.* Glencoe, IL: Free Press, 1958.
———. *Economy and Society.* New York: Bedminster Press, 1968.

———. *The Theory of Social and Economic Organization.* Edited by Talcott Parsons. London: Free Press, Macmillan Publishing Co., 1964.
White, William. *City: Rediscovering the Center.* New York: Anchor Press, 1988.
Wu Hung. *Monumentality in Early Chinese Art and Architecture.* Stanford: Stanford University Press, 1995.
Zukin, Sharon. *The Cultures of Cities.* New York: Wiley, 1995.

INDEX